The Urban Heat Island

The Urban Heat Island

A Guidebook

Iain D. Stewart
Gerald Mills

ELSEVIER

Elsevier
Radarweg 29, PO Box 211, 1000 AE Amsterdam, Netherlands
The Boulevard, Langford Lane, Kidlington, Oxford OX5 1GB, United Kingdom
50 Hampshire Street, 5th Floor, Cambridge, MA 02139, United States

Notices
Knowledge and best practice in this field are constantly changing. As new research and experience
broaden our understanding, changes in research methods, professional practices, or medical treatment
may become necessary.

Practitioners and researchers must always rely on their own experience and knowledge in evaluating
and using any information, methods, compounds, or experiments described herein. In using such
information or methods they should be mindful of their own safety and the safety of others, including
parties for whom they have a professional responsibility.

To the fullest extent of the law, neither the Publisher nor the authors, contributors, or editors, assume
any liability for any injury and/or damage to persons or property as a matter of products liability,
negligence or otherwise, or from any use or operation of any methods, products, instructions, or ideas
contained in the material herein.

Library of Congress Cataloging-in-Publication Data
A catalog record for this book is available from the Library of Congress

British Library Cataloguing-in-Publication Data
A catalogue record for this book is available from the British Library

ISBN: 978-0-12-815017-7

For information on all Elsevier publications
visit our website at https://www.elsevier.com/books-and-journals

Publisher: Janco, Candice (ELS-HBE)
Acquisitions Editor: LaFleur, Marisa (ELS-CMA)
Editorial Project Manager: Gammell, Ruby (ELS-CON)
Production Project Manager: Vignesh Tamil
Cover Designer: Hitchen, Miles (ELS-OXF)

Typeset by SPi Global

Working together
to grow libraries in
developing countries

www.elsevier.com • www.bookaid.org

Contents

List of figures and tables

List of figures

List of tables

Introduction

The urban heat island, in all of its manifestations, is a ubiquitous outcome of urbanization. The term *urbanization* is used to describe two distinct, but related, processes. First, it describes the absolute and relative proportion of the population living in densely settled spaces and engaged mostly in nonagricultural activities. Second, it describes the radical transformation of the natural landscape (e.g., the flattening of topography, the modification of stream channels, and removal of vegetation) to convert it to one that is suited to human habitation. The combination of dense living (with its concomitant energy and material needs) and manufactured spaces creates a distinctive urban climate.

The simplest definition of the urban heat island (UHI) is that it represents a difference in the equivalent temperatures of the city (and its parts) and the surrounding natural (nonurbanized) area. This is based on the premise that the natural landscape represents the temperature where the city is located if there were no urbanization. There are four distinct types of UHI (Fig. 1.1):

1. The canopy-level UHI (CUHI) is based on the near-surface air temperature measured below roof height;
2. The boundary-level UHI (BUHI) is based on air temperature measured well above the height of buildings in cities;
3. The surface UHI (SUHI) is based on the temperature of the three-dimensional urban surface, that is, the ground, walls, and rooftops;
4. The substrate UHI (GUHI) is based on the temperature of the soil below the ground surface.

Logically, the urban temperature effect begins at the surface (SUHI), where the crenulated and manufactured envelope seals the ground and encloses the indoor building space. The distribution and absorption of solar energy at this surface and its subsequent transfer into the atmosphere and substrate results in marked temperature variations at the microscale. Added to these natural exchanges is the injection of heat energy into the air and the substrate through the exhausts of cars, buildings, pipes, etc. The urban effect on the soil and geology beneath the city (GUHI) has received little attention apart from places where either the impact is visible (e.g., melting permafrost in Arctic villages) or preexisting measurements for an unrelated study are available.

The most common UHI study examines the urban effect on the near-surface air temperature ($\sim 2\,\text{m}$ above the ground), which in cities places the instrument within the roughness sublayer (RSL) of the UBL and specifically within the urban canopy layer (Fig. 1.1). Here, the sensor responds to its immediate environment including nearby walls, ground, gardens, etc. Elevating the thermometer changes its exposure: above roof level, the sensor records the contributions of an ever-increasing area incorporating the contributions of rooftops, walls, streets, carparks, trees, etc. At this height, individual contributions are difficult to identify owing to the turbulent nature of the RSL and there can be no guarantee that temperature measurements are representative of

The Urban Heat Island. https://doi.org/10.1016/B978-0-12-815017-7.00001-1

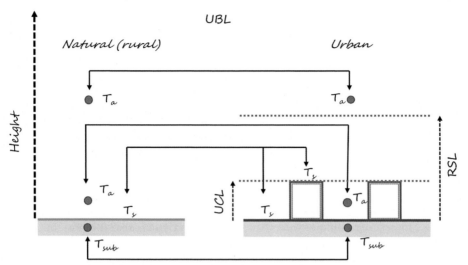

Fig. 1.1 Types of urban heat island (UHI). The magnitude of each type is assessed by comparing temperatures in the city against a benchmark, typically a natural (rural) environment. The surface UHI compares temperatures at the solid-air interface (T_s); the canopy-level UHI compares near-surface (2 m) air temperatures (T_a); the boundary-layer UHI compares air temperatures well above the underlying surface (T_a); and the substrate UHI compares temperatures below the surface (T_{sub}). The abbreviations refer to the urban canopy layer (UCL), the roughness sublayer (RSL) and the urban boundary layer (UBL).

the underlying urban landscape. Raising the sensor above the RSL places it within the inertial sublayer (ISL), where the diverse contributions of the underlying landscape are thoroughly mixed (Fig. 1.2). Located well above the UCL (2–4 times the heights of buildings), the measurements will represent the base of the deeper urban boundary layer (UBL), effectively the "surface" as far as the UBL is concerned. Above this sublayer, the BUHI can be measured within the mixed layer but the cost of measuring at this height using very tall masts or airborne platforms (e.g., balloons and aircraft) means that there are few studies at these levels.

1.1   A brief history of UHI studies

The near-surface temperature effect has been studied for over 200 years and has generated an immense literature that is extremely diverse in terms of content, spatial coverage, methodological approach, and experimental rigor.

The origin of urban heat island science is the work of Luke Howard on the *Climate of London* (the first edition of which was published in 1818). Over a period of 26 years, he and his family made daily measurements of maximum and minimum air temperature at different places outside the city, with a view to describing the climate of the

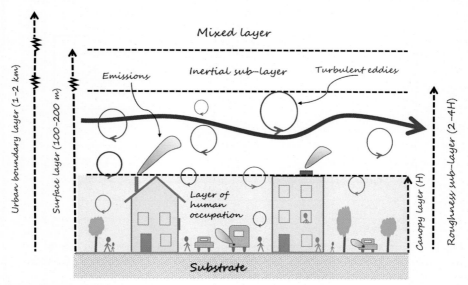

Fig. 1.2 The structure of the lowest part of the urban boundary layer (UBL). The UBL (1–2 km in depth) is comprised of a mixed layer and a surface layer (100–200 m deep). The latter includes the inertial and roughness sublayers and the urban canopy layer (UCL), which describes the space between the ground and the building rooftops.

place where London is situated. In evaluating his work, he compared his records with those made in the city by the Royal Society (the preeminent scientific body of the day) and discovered a systematic difference that he could not attribute to observational errors. He concluded that *the temperature of the city is not to be considered as that of the climate* (Fig. 1.3) (Howard, 1833). His work found that the differences were greatest in the winter months when the city was warmer, and he hypothesized that this difference was due to anthropogenic heating of buildings, the lack of vegetation to cool air, and obstructions to the ventilation of urban air. This simple comparison of near-surface air temperatures—one recorded at a countryside site (often described as "rural") and the other at an urban site in the center of the city (ΔT_{u-r})—is an established methodology for the CUHI assessment that is still used.

Throughout the nineteenth century and much of the twentieth century, the study of heat islands was largely the concern of climatologists interested in microscale climate changes. Much of the early work was conducted in mid-latitude cities of Western Europe and Japan, but after 1945 such studies became much more common. The observational evidence gradually became more sophisticated as the ability to measure and record precise air temperatures improved and techniques were developed to detect the spatial pattern of a heat island. Fig. 1.4 shows the London UHI as illustrated in Chandler's (1965) *Climate of London*. Note that the map shows the minimum air temperature at night during calm weather with clear-sky conditions. The CUHI is revealed by the alignment of the isotherms with the urban footprint and the increasing

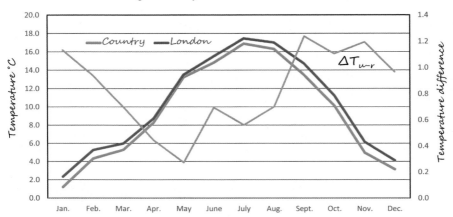

Fig. 1.3 The mean monthly air temperature (°C) in London and in the "country" for the period 1807 to 1816 (left-hand axis). The UHI magnitude (ΔT_{u-r}) is the difference between London and county temperatures (right-hand axis).
Based on Howard, L., 1833. The Climate of London. Harvey and Darton, London. Available at www.urban-climate.org.

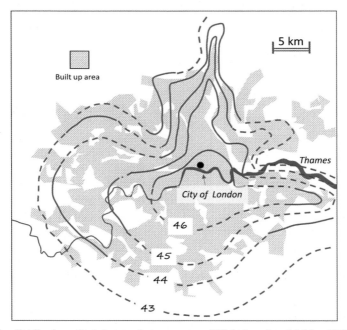

Fig. 1.4 The distribution of minimum air temperature (°F) in London, 14 May 1959 (based on Fig. 55 in Chandler, 1965). Dashed-line isotherms indicate areas of uncertainty. Weather conditions: light northeasterly to northerly winds of less than $2\,\mathrm{m\,s}^{-1}$ and clear skies associated with a deep anticyclone. The location of the City of London, where the Royal Society temperature readings used by Howard was based, is labeled.

values toward the city center. By the mid-1970s, it was common knowledge among climatologists that all cities (and settlements) create a CUHI of varying magnitude and extent that is modulated by the background weather. The CUHI was confirmed to be greatest at night, under clear and calm conditions when the canopy-level air cooled more slowly than the near-surface air outside the city.

During the 1970s, measurements began of the air above the city, often as part of wider studies into air quality and the transport of pollutants downwind from urban areas. Using airborne platforms, observations within the mixed layer of the UBL revealed that the warming effect of cities extended to a height of 1–2 km in the daytime. While this phenomenon had some common features to the near-surface UHI, it also had unique features: for example, at this level, the warming influence of the city was present by day and night. Clearly, the boundary-layer and the canopy-layer UHIs were different. Within the UCL, microscale processes dominate and the role of building walls becomes increasingly important (Fig. 1.2), but well above the rooftops, atmospheric mixing blends the contributions of the diverse urban landscapes. This distinction is a very important conceptual step in structuring further scientific research on the UHI into that within the urban canopy layer (CUHI) and that in the urban boundary layer (BUHI).

Prior to the 1970s there were few observations of surface temperature in cities largely because thermal infrared (TIR) sensors were expensive and difficult to deploy. Large-scale assessment of the urban surface required that these instruments were sufficiently elevated so that the instrument could "see" a swath of the urban landscape and aggregate the contributions of the ground, walls, and rooftops. Satellite observation systems since the mid-1970s have included TIR sensors that estimate land surface temperature (LST) and permit assessment of the surface UHI. Rao (1972) identified the potential of satellite-based TIR systems to provide *a means of monitoring the growth and development of urban and suburban areas and their impact on the environment* (Fig. 1.5). Since then, the accumulation of observations and the improvement in satellite technology has generated an immense body of knowledge on the SUHI that is global in extent. Even still, the satellite perspective alone cannot resolve many of the processes responsible for the SUHI and this work must be complemented by on-the-ground studies.

Historically, the study of CUHI has focused on nighttime near-surface air temperature patterns that result from differential cooling, whereas SUHI studies have focused on the diurnal pattern in observed surface temperatures that results from differential heating. Unlike the nighttime CUHI, which displays a coherent spatial pattern that corresponds with the urban footprint and locates the highest values in the densely built city center, the daytime SUHI has a more complex pattern. While higher temperatures are strongly correlated with impervious surface cover, multiple hot spots are often associated with the specific make-up of different urban areas. At night, the SUHI (as seen by TIR sensors) has the same basic pattern as the nighttime CUHI but the intra-urban temperature differences are often smaller. Linking the surface and air temperatures has been described as *the greatest unknown in remotely sensed studies of heat islands* (Nichol et al., 2009), and research into this topic is an area of active work.

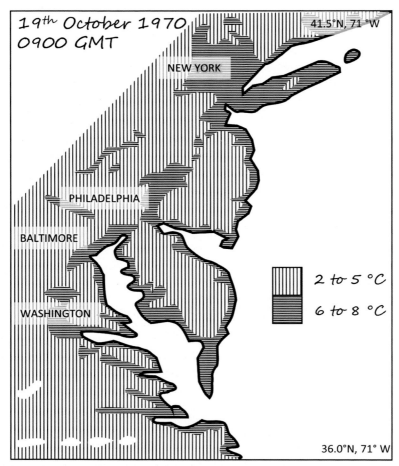

Fig. 1.5 Analysis of digitized infrared data from the Improved TIROS Operational Satellite (ITOS-1) for 19 October 1970 at 0300 h local time. Heavy shading represents regions with surface temperatures of 6–8 °C; vertically hatched areas, 2–5 °C.
Source: Rao, P.K., 1972. Remote sensing of urban "heat islands" from an environmental satellite. Bull. Am. Meteorol. Soc. 53, 647–648, with permission.

1.2   Why continue to observe the UHI?

A typical description of scientific progress in a subject area, such as the UHI, outlines a process of discovery, observation, analysis, testing, and confirmation that results in a thorough understanding of its causes, which is universal in its application. Once this sequence is complete there is often little scientific merit in further study unless the fundamental premise of the explanation has been called into question. In some respects, the study of UHI fits this sequence.

While there were visual observations of the urban warming effect prior to Howard's study of the CUHI, his was the first to measure it. Subsequent studies in many cities and towns confirmed the universality of the phenomenon but each place had its own unique CUHI depending on the nature of the study and the character of the urbanized landscape, that is, the local topography and urban layout. Overtime, the accumulation of data on the CUHI permitted a search for common causes by establishing controls on atmospheric conditions and isolating individual urban characteristics for examination. The shift in the 1980s toward a process-based understanding moved the research frontier from simply the measurement and comparison of air temperatures to an exploration of the energy exchanges that underpin the formation of the CUHI. Similarly, the study of SUHI, which is closely linked to satellite-based TIR instruments, has advanced considerably over the last few decades, as data have accumulated and instruments have become progressively more sophisticated in terms of spectral and spatial resolution. So, the question arises, given the history of UHI studies, what can new research contribute?

Leaving aside the substrate and boundary-layer UHIs where there is still an obvious need for observational data, let us focus on three reasons for continuing to study the surface and canopy-level UHIs.

1.2.1 UHI case studies

The measurement of UHI patterns in individual cities can provide valuable datasets for other studies, while guiding urban planning and design policies and filling in gaps in the existing descriptive literature. These gaps are linked to coverage rather than content (with a few exceptions), as most studies have been completed for cities of the mid-latitudes with distinct seasonal patterns that affect heating/cooling needs and vegetative growth. A great many of these cities have followed common patterns of urban development, featuring low-density outskirts with considerable green coverage and a densely built urban core. As a generalization, there are few UHI studies that have been completed in the following settings:

- Tropical and Arctic regions where the background climate-drivers (e.g., wind, sunshine, rainfall, and pressure) and natural land-cover characteristics are different from those of the mid-latitudes;
- Cities that do not follow the urban layout described above, such as places with extremely tall, closely spaced buildings that may be located in the city outskirts, rather than in the center;
- Cities that are expanding (horizontally and vertically) very rapidly and where benchmark UHI values can be used to evaluate the effect of urban land management policies;
- Informal settlements where the material properties of building and roads may be highly variable and assumptions based on formal settlements may be incorrect;
- Highly complex topographies where the urban influence must be isolated from the panoply of other effects.

In all cases, high-quality observational data on the UHI will contribute to the field if the researcher follows methodological standards that allow comparison with other studies.

1.2.2   UHI impacts

The near-surface air and surface temperatures can be regarded as a simple diagnostic tool to identify where and when the urban warming impact is greatest, and to guide planning and design interventions. Temperature can be described as a "response" variable that represents the net impact of a series of interacting exchange processes, most of which are difficult to measure. Observing the UHI then can be used to guide further research that examines the site-specific causes, such as surface albedo and vegetation cover, that result in hot/cool areas in the city. In the same vein, temperature measurements have been used to evaluate atmospheric models that simulate all of the processes that govern the urban climate; if simulations can reproduce the magnitude and timing of the UHI, then this provides strong support for the model simulations, which can be used to evaluate impacts.

Responses to climate changes are often categorized into those designed to mitigate and to adapt. The former seek to identify and offset the causes of the UHI to reduce or eliminate its magnitude, while the latter adjusts systems and infrastructure to cope with higher temperatures. In reality, most policies include both perspectives and the emphasis depends a great deal on the impacts of the UHI. Higher urban temperatures are associated with increased energy use for cooling buildings, enhanced heat stress on humans, and changes to natural ecosystems. Moreover, the UHI occurs alongside other urban effects on air pollution, airflow, hydrology, etc., so that actions taken to mitigate the UHI can improve the urban environment generally. However, the UHI can have some positive outcomes for some cities. Writing on the cool mid-latitude climate experienced in London, Chandler (1965) suggested that *the higher autumn, winter and spring nighttime temperatures which reduce heating costs and lengthen the frost-free period* could be considered favorable. Elsewhere, in warmer climates, adaptation to higher temperatures would require cooler outdoor and indoor spaces to offer respite during hot weather, and modifying buildings and infrastructure to adjust to the changed climate. All this to say that decisions on managing the UHI, and designing an appropriate response strategy, should evaluate its spatial and temporal impacts and its correlation with other undesirable urban effects.

1.2.3   Climate change interactions

Anthropogenic climate change occurs at varying degrees across all climatic scales as a consequence of landscape changes and emissions of energy, gases, and particulates. At the global scale, increased concentration of greenhouse gases (GHG) in the atmosphere and the ocean is the primary driver of climate change, although widespread modification of natural land cover is a contributing factor. The primary source of GHGs is the combustion of fossil fuels (coal, gas, and oil) which generates carbon dioxide (CO_2). The major drivers of anthropogenic CO_2 emissions are cities, especially those of wealthy economies. The global accumulation of emissions is causing changes to weather and climate patterns that are manifest as sea-level rise, atmospheric warming, and more extreme weather events, such as heatwaves. Cities are at particular risk from these changes for a number of reasons. First, cities are located in common topographic settings—close to sea level, in valleys and basins, and near coasts—that expose them to a range of hazards. Second, the high concentration of population and the dense urban infrastructure increases their vulnerability to the projected changes in climate. Finally, the urban effect on local climate, such as the UHI, will enhance expected large-scale climate changes like global warming and heatwaves.

Here it is worth distinguishing between urban scale and global scale processes and outcomes. First, although GHG concentrations in urban areas are higher than current and predicted global concentrations, this is not a cause of the UHI. Second, the extent of urbanized landscape globally is relatively small (1–5%, depending on the definition), and while these changes to the landscape are profound, the wider impact is limited by comparison with the historic global land-cover changes associated with forest clearance and agricultural production. Finally, the UHI effect has no significant bearing on the average global temperature, except insofar as it may bias observations at meteorological stations that are poorly sited for their intended use. So, the focus on cities in global climate change is largely to mitigate GHG emissions (that is, reduce fossil-fuel energy use) and to consider how best to adapt to projected climate changes.

Given this context, observing the UHI is relevant to global climate studies for a number of reasons:

- To "correct" station observations that are used to assess global and regional climate change by removing the urban temperature effect;
- To evaluate the net impact of both global climate changes and the UHI on urban residents and the environment;
- To design mitigation/adaptation policies that address climate change across the spatial hierarchy (that is, urban, regional, and global scales);
- To represent the projected climate of cities in the future.

Much of the evidence for global climate change is based on temperature measurements made at conventional weather stations, many of which have lengthy records that predate other sources, such as satellite observations. According to the World Meteorological Organization (WMO, 2008), the *best site for the measurements is over level ground, freely exposed to sunshine and wind and not shielded by, or close to, trees, buildings and other obstructions.* At most standard weather stations, the ground cover is short grass and the thermometer is housed in a ventilated radiation shield 2 m above the surface. Stations that follow these guidelines can provide observations that respond to regional weather and climate patterns. In using these types of data to estimate regional climate records and assess global near-surface air temperature, one must overcome many problems (e.g., uneven sampling, site re-location, and instrument changes), among which is the potential for urban influences. In this respect, the UHI effect is seen as a potential contaminant that should be removed. Much research has focused on evaluating the quality of observations across a network of stations and employing simple measures of land-cover/population change to assess the magnitude of the urban influence on recorded air temperature.

From the perspective of future urban climates, the implication of global climate change (GCC) is that the background climate experienced by cities will change and that the urban contribution can enhance/diminish undesirable outcomes.

The interaction of global warming and the UHI is of particular concern in the literature because the projected temperature changes (timing and magnitude) align closely with the UHI climatology. For many places this will bring increased heat stress and impose a burden on people, energy systems, etc. Ideally, then, mitigation strategies that seek to address GCC by modifying urban form and function should necessarily consider the UHI implications. For example, compact cities (often defined as high-density spaces with tall, closely packed buildings) may reduce overall energy use and reduce GHG emissions, but could also enhance the UHI and create heat stress if done inappropriately. UHI studies are needed to develop policies that account for the hierarchy of climatic scales and consider the impacts at those scales.

Finally, it is worth pointing out that some research has used the city as an analogy to examine the influence of GCC on ecosystems. This type of work assumes that many of the outcomes of GCC, including surface and atmospheric warming, are already present in cities and that much can be learned of ecosystem responses in the future by examining current responses in urban environments.

1.3   The purpose of this book

This book is written for novices in the field of urban climatology for whom the UHI phenomenon is their first direct encounter with climate research, and observing/measuring its environmental impact is part of a course project or a larger study. However, while the field has a long history, there are currently no guidelines to help conduct a heat island study. As a result, a great deal of confusion exists in the literature about the types of UHI and their measurement approaches. Unfortunately, this inhibits productive knowledge transfer and effective policy formulation.

The main purpose of this book is to provide guidelines to plan and execute an observational study of the urban temperature effect; to analyze and interpret the data collected from the field; and to communicate the results to their appropriate audiences. Its focus is the canopy-level and surface UHIs that together make up the great majority of heat island studies. The book is divided into two parts: the first part outlines the physical processes responsible for the UHI and common actions used to manage it; the second part presents methodological guidelines for CUHI and SUHI studies. Throughout, we have limited our use of references, as much of the material has moved from "new" to "established." At the end of the book, we provide a list of the most relevant publications from our vantage point.

References

Chandler, T.J., 1965. The Climate of London. Hutchinson, London. Available at www.urban-climate.org.
Howard, L., 1833. The Climate of London. Harvey and Darton, London. Available at www.urban-climate.org.

Nichol, J.E., Fung, W.Y., Lam, K.S., Wong, M.S., 2009. Urban heat island diagnosis using ASTER satellite images and 'in situ' air temperature. Atmos. Res. 94, 276–284.

Rao, P.K., 1972. Remote sensing of urban "heat islands" from an environmental satellite. Bull. Am. Meteorol. Soc. 53, 647–648.

WMO (Ed.), 2008. Guide to Meteorological Instruments and Methods of Observation, seventh ed. World Meteorological Organization, Geneva. WMO-No. 8.

Part One

The first part of this book establishes the physical basis for the surface and canopy-level UHIs, which sets the context for a discussion on UHI mitigation and/or adaptation. There are a great many scientific papers that cover the substance of Part One and our intent here is to describe the overall findings of this work in words and symbols. Part Two of the book uses this knowledge to present guidelines for performing UHI research.

In Chapter 2, we use the energy budget to provide theoretical context for describing the response of surface, substrate, and air temperatures to urban landscape change. Fundamental to this approach is the principle of conservation of energy, which simply states that energy can neither be created nor destroyed, but can change its form. To apply this principle, we track the exchanges and stores of energy in the atmosphere and substrate close to the ground. In this perspective, temperature change is a response to physical processes that regulate the exchange of energy (termed a *flux*). These processes include the diurnal (and seasonal) variations in Earth-Sun geometry, surface topography, and weather patterns. We use symbols to represent important process and response variables, such as the sensible heat flux (Q_H) and air temperature (T_a), respectively. Apart from budget statements that require adding and subtracting, we link relevant variables by stating that the magnitude and direction of change of one is a function of (*f*) other variables, without providing explicit details of this relation.

In Chapter 3, we examine the policy options to manage the UHI given the physical causes described in Chapter 2. Climate-based policies are often categorized into those designed for mitigation and adaptation, but there is considerable overlap when either is applied to the UHI. Here, we use *mitigation* to refer to modifications of the paved and built land-cover that are designed to reduce excess urban heat. These modifications can be complemented by measures to reduce anthropogenic energy generation. We use *adaptation* to describe society's adjustment to the heat conditions that already exist (or that will exist) due to urbanization.

The energetic basis

2

The UHI is a result of urbanization, which brings both landscape change (land cover) and intense occupation (land use) to an area. Land-cover change refers to the removal of natural vegetation, sealing soil under paving, and constructing buildings. Land use refers to the occupation of space that drives the energy and material flows (fuel, water, food, etc.), which are needed to maintain the urban economy, heat/cool houses and offices, sustain transport systems, and so on. These flows generate wastes that are deposited into the atmosphere, the soil, and the water—both inside and outside the city limits. The consequences of these changes include the alteration of surface, subsurface, and air temperatures, but it is important to recognize that temperature is a response to a set of energy exchange and storage processes.

In this chapter, we outline energy exchanges at the Earth's surface (Section 2.1) and describe the magnitudes and patterns of these exchanges over different surface types (Section 2.2). We then describe the make-up of the urban landscape and how this results in distinct local and microscale outcomes (Section 2.3). Lastly, we discuss the types of UHI within an energy budget framework that accounts for spatial and temporal scales.

Throughout the chapter, symbols are used to represent these energy fluxes and their relation with other variables, which may be more recognizable, such as wind speed (v) or air temperature (T_a). These fluxes are often expressed as a function of (f) the variables; for example, the longwave radiation emitted by an object ($L\uparrow$) is a function of its surface temperature (T_s) and emissivity (ε_s), so $L\uparrow=f(T_s,\varepsilon_s)$. While this approach limits a fuller understanding of the underlying physics that is available in many texts, hopefully it does not distract from understanding the principles that underpin the UHI phenomenon.

2.1 Energy and energy transfer

For the topics discussed in this book, there are three relevant forms of energy (radiation, sensible, and latent heat) and three modes of transfer (radiation, convection, and conduction). The scientific units associated with energy exchanges are the following: Joules (J) for energy; Watts (W) for the rate of energy flow (that is, a Joule per second, $J\,s^{-1}$); and Watts per square meter ($W\,m^{-2}$) for the rate of energy exchange across a plane, also termed a *flux density*.

2.1.1 Radiation

Radiation is both a form of energy and a mode of energy transfer. Radiation is emitted by all objects with T_s above $-273\,°C$ (or 0 Kelvin). This energy travels at the speed of light ($300\times10^6\,m\,s^{-1}$) as a series of waves and does not require a medium, so it can pass through the vacuum of Space. The intensity of radiation energy is inversely related to

The Urban Heat Island. https://doi.org/10.1016/B978-0-12-815017-7.00002-3

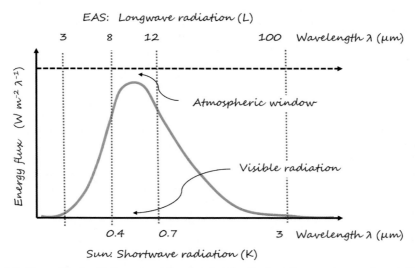

Fig. 2.1 Generalized radiation curves showing the distribution of energy by wavelength for the Sun (bottom axis) and the Earth-atmosphere system (EAS) (top axis).

its wavelength (λ), and an object emits radiation across a spectrum of wavelengths. A plot of the radiation flux density ($E\lambda$) emitted per wavelength ($W\,m^{-2}\,\lambda^{-1}$) describes a distinctive curve shape with a single peak (λ_{max}) and a positive skew (Fig. 2.1). The total energy emitted ($W\,m^{-2}$) is proportional to the emissivity (ε) and temperature (in Kelvin) of the object:

$$E = f\left(\varepsilon, T^4\right) \tag{2.1}$$

Emissivity (ε), which indicates the efficiency (0–1) of the emitting object is wavelength dependent but, for our purposes, we simply use a bulk emissivity value to represent the efficiency across a range of wavelengths. In climatology, there is a clear distinction between radiation from the Sun, and from sources within the Earth-atmosphere system (EAS) that have very different temperatures and consequently different emission spectra.

- Shortwave (K) radiation is emitted by the Sun, which has a surface temperature of about 6000 Kelvin. At the top of the atmosphere the spectrum of solar radiation ranges from 0.3 to 3 μm and has a peak in the visible radiation band from 0.4 to 0.7 μm. Given the distance of the Earth from the Sun and the absence of any intervening material, solar radiation is received as a beam with an origin and a direction. On a surface perpendicular to the solar beam outside the atmosphere, the shortwave radiation received is about 1370 W m^{-2} (referred to as the solar constant); this is a useful number to bear in mind, as the amount received at the ground will always be lower!
- Longwave (L) radiation is emitted within the EAS and from the EAS into Space. The average temperature of the EAS is about 290 K and the emission spectrum ranges between 3 and 100 μm, with a peak between 8 and 14 μm; this spectrum is often termed *far infrared* or *thermal radiation*. Longwave radiation is emitted diffusely in all directions.

Upon encountering a medium, radiation can be transmitted (pass through unaltered), absorbed (stored in the medium), or scattered (redirected). The fate of energy of a given wavelength depends on how it interacts with the contents of the medium. Absorption results in heating, and the ability to absorb (absorptivity) is equal to the ability to emit (emissivity) at that wavelength. For an opaque solid there can be no transmission and all scattering occurs away from its surface as reflection; for these objects, reflectivity is the inverse of emissivity. For gases and liquids, scattering can occur in all directions, forward and backward. While clouds scattering visible light in the direction of the observer appear white, the clear sky scatters blue light. Table 2.1 lists the radiative properties of various materials. Short-wave (solar) reflectivity is termed *albedo* (α) and we reserve the term *emissivity* (ε) for longwave (terrestrial) radiation, as the Earth is not a source of shortwave radiation emission.

2.1.1.1 Radiation sources and geometry

Direct (beam) radiation is sourced from the Sun and has a direction (Fig. 2.2A) associated with its position—given by the solar azimuth (angle from north) and zenith (angle from perpendicular)—which changes with time of day and time of year. The amount of direct radiation incident on a surface (S) depends on the respective geometries of the Sun and the surface and is modulated by transmissivity of the atmosphere. Diffuse solar radiation is sourced from an area and does not generate shadows. As the solar beam enters the atmosphere it will be scattered at a rate that depends on the composition of the atmosphere and the distance it must travel. For a clean and clear atmosphere, about 70% of the solar energy available outside the atmosphere will be transmitted but this

Table 2.1 Radiative properties of natural and manufactured materials.

Surfaces	Albedo (α)	Emissivity (ε)
Grass	0.16–0.26	0.90–0.98
Forest	0.13–0.20	0.90–0.99
Water	0.03–0.10	0.92–0.97
Desert sand	0.20–0.45	0.84–0.92
Snow	0.50–0.90	0.82–0.99
Asphalt	0.05–0.27	0.89–0.96
Brick	0.2–0.6	0.90–0.92
Clear glass	0.08	0.87–0.95
Concrete	0.10–0.35	0.85–0.97
Tile	0.01–0.35	0.90
Tar and gravel	0.08–0.18	0.92
White paint	0.50–0.90	0.85–0.95
Corrugated iron	0.10–0.16	0.13–0.28

The values for water and glass are dependent on the angle of the solar beam at the surface; those for natural cover are dependent on seasonal growth; and that for snow depends on its condition (e.g., dry, wet). The properties of manufactured materials change with age and become darker (lighter) with time.
Source: Oke, T.R., Mills, G., Christen, A., Voogt, J.A., 2017. Urban Climates. Cambridge University Press, Cambridge, UK.

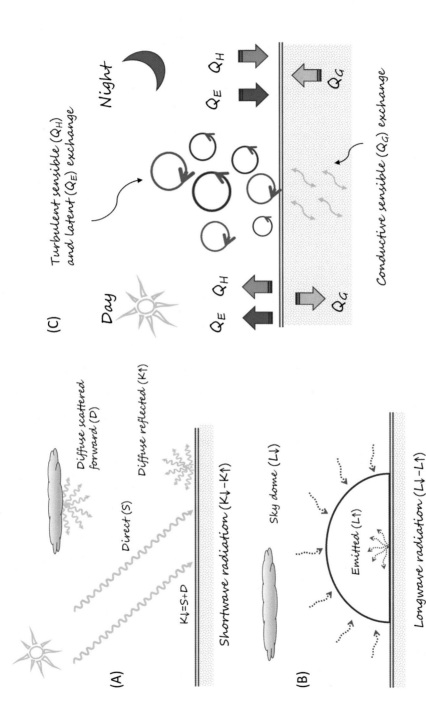

Fig. 2.2 (A) Shortwave and (B) longwave radiation exchanges at the ground, and (C) nonradiative exchanges by turbulence with the atmosphere and by conduction with the substrate.

decreases considerably with poor air quality and low cloud cover. The atmosphere is a relatively poor absorber of shortwave radiation, so much of the remainder is scattered toward the Earth's surface as diffuse radiation. As a simplification, this diffuse solar radiation receipt at a surface (D) can be treated as originating from a uniform sky dome. The total shortwave radiation receipt at the surface (K↓) is the sum of the direct and diffuse sources (S + D). Reflected shortwave radiation at the Earth's surface (K↑) is scattered and for most surfaces it can be treated as diffuse solar radiation, but there are exceptions. Glass, for example, can redirect beam radiation if the angle of receipt is small. Longwave radiation exchanges are diffuse in character (Fig. 2.2B). Longwave radiation received at the surface (L↓) is acquired from the overlying atmosphere and can be treated as arriving from the sky dome. Longwave radiation emitted at the surface (L↑) can similarly be visualized as entering a hemisphere.

The distinction between beam and diffuse radiation is relevant in assessing the sources and magnitudes of radiation receipt and loss. Direct solar radiation (S) receipt is regulated by the relative geometry of the Sun and of the surface (slope and aspect angles), as well as the transmissivity of the atmosphere and the presence of intervening obstructions. Diffuse radiation sources (D and L↓) originate from a hemispheric dome. All of the individual features that reflect or emit toward the surface of interest can be "mapped" onto this dome. Each of the features occupies a fraction of the dome area (2π steradians) that is termed a *view factor*; a simple division of view factors is into the sky view factor (SVF) and its reciprocal (1-SVF). Just as this map represents sources of diffuse radiation, it also shows the fate of radiation emitted or reflected from the surface of interest.

2.1.2 Sensible heat

Sensible heat describes the energy stored in a volume and is linked to temperature. It is associated with the microscopic motion of the molecules that constitute the material and can be sensed by humans. The property that links temperature change to energy content is heat capacity (C); the specific heat capacity has units of Joules per kilogram per Kelvin ($J\,kg^{-1}\,K^{-1}$). There are 1.25 kg of air per cubic meter of atmosphere at sea level. Table 2.2 lists the thermal properties of a selection of natural and manufactured materials and, in the case of the former, their response to water content.

Sensible heat energy is transferred along a temperature gradient (that is, from a higher to lower energy state) by conduction and by convection (Fig. 2.2C).

- Conduction occurs by molecular action and is slow; the pace of sensible energy transfer through a medium depends on its conductivity (k), expressed as Watts per meter per second ($W\,m^{-1}\,s^{-1}$). In solids, the transfer of sensible heat by conduction (Q_G) dominates.
- Convection describes exchanges that occur as bodies of air (and their properties) mix. We reserve the term *convection* to describe vertical exchanges of sensible heat (Q_H) and use *advection* to describe horizontal exchanges.

Keep in mind that the exchange of sensible heat only takes place when there is a temperature gradient; if there are no vertical/horizontal differences in temperature, then there is no convection/advection of sensible heat, even if it is windy.

Table 2.2 Thermal properties of natural and manufactured materials.

Materials	Heat capacity (C) $(MJ\,m^{-3}\,K^{-1})$	Thermal conductivity (k) $(W\,m^{-1}\,K^{-1})$	Thermal admittance (μ) $(J\,m^{-2}\,s^{-\frac{1}{2}}\,K^{-1})$
Clay soil			
• Dry	1.42	0.25	600
• Saturated	3.10	1.58	2210
Sandy soil			620
• Dry	1.28	0.3	2550
• Saturated	2.96	2.2	
Water (4 °C)	4.18	0.57	1545
Air (10 °C)	0.0012	0.025	5
• Still	0.0012	125	390
• Turbulent			
Asphalt	1.94	0.75	1205
Brick	1.37	0.83	1065
Clay tiles	1.77	0.84	1220
Glass	1.66	0.74	1110
Concrete			
• Aerated	0.28	0.08	150
• Dense	2.11	1.51	1765

Source: Oke, T.R., Mills, G., Christen, A., Voogt, J.A., 2017. Urban Climates. Cambridge University Press, Cambridge, UK.

2.1.3 Latent heat

Latent heat is the energy associated with the physical state of water, which can readily change its form from solid, to liquid, to vapor, and vice versa. The transformation to a higher energy state requires breaking apart the bonds that connect the molecules; it requires 0.33 MJ to change 1 kg of ice into water (melting), and 2.56 MJ to change 1 kg of water to vapor (evaporation). Although this energy is stored in the volume, it does not reveal itself as a temperature change and it is not sensed by humans. Water vapor represents the highest energy state for water and it is measured in many different ways, including vapor density (ρ_v in kg m^{-3}), vapor pressure, and mixing ratio. The maximum (or saturation) vapor content of air increases with temperature. The relative humidity (RH) compares the actual to the saturation vapor content of a volume and is a useful measure of the potential for evaporation from a surface. Critically, humans are sensitive to RH rather than vapor content. Water in air with a RH of close to 100% will condense into droplets and the latent heat of evaporation is released as sensible heat (warming). Finally, latent heat is transferred in the atmosphere by convection (Q_E), just like Q_H (Fig. 2.2C).

2.2 Energy balances and budgets

A surface in climatic terms represents an interface that divides what is on either side. By definition, it cannot store energy but is a plane across which energy is exchanged and the sum of these exchanges must be zero. A volume is enclosed by a surface and

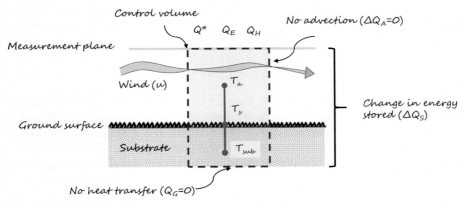

Fig. 2.3 Measurement points for air (T_a), surface (T_s), and substrate (T_{sub}) temperatures at a grass-covered observation site. The dashed area is a conceptual "control volume" that extends into the atmosphere and into the substrate to a depth where the diurnal temperature change is negligible. If the site is over a flat, extensive, and homogenous surface, then advection is negligible. In these circumstances, measuring net radiation (Q^*) and the turbulent sensible (Q_H) and latent (Q_E) heat exchange at an elevated measurement plane allows assessment of the heat stored in the volume of air and soil (ΔQ_S), which is revealed in temperature changes.

can be defined to represent an object of interest with a clear boundary (e.g., a tree, person, or building), but it could also be used to enclose a volume of interest, such as a near-surface layer of atmosphere or a depth of soil. For the volume that is contained by this surface, the aggregate exchange across all bounding surfaces (that is, the budget) need not be zero: a positive value indicates an increase in energy content (convergence) due to the accumulation and storage of energy and vice versa (divergence). For clarity, we will use the term *surface* for the interface between two media, such as the soil substrate and the atmosphere where different exchange processes dominate. We use the term *plane* to describe a level that divides the same medium, such as the height (or depth) that represents a convenient platform to make measurements (Fig. 2.3).

To establish the energetic context for the UHI, we will present energy balances and budgets for a series of microscale environments to represent a natural surface, a paved surface, and a building.

2.2.1 An extensive short grass surface

Assume that this is a homogenous, level-ground surface with a covering of short grass. It is the type of surface above which a standard weather station is situated, and is often used as a benchmark against which the urban temperature effect is measured. The sum of all incoming ($\downarrow$) and outgoing ($\uparrow$) radiation at a surface is termed *net radiation* (Q^*):

$$Q^* = K\downarrow - K\uparrow + L\downarrow - L\uparrow \tag{2.2}$$

Shortwave radiation ($K\downarrow$) is received during the daytime, and under clear skies the magnitude peaks at noon, corresponding to the highest elevation of the Sun above the horizon. It consists of direct (S) and diffuse (D) parts:

$$K\downarrow = S + D \tag{2.3}$$

As this surface is extensive and flat, there are no sizeable vertical elements that can cast shadows or block part of the sky hemisphere. In these circumstances, diffuse radiation is sourced from the entire sky dome and the SVF of the surface equals unity. $K\uparrow$ is reflected shortwave radiation and is regulated by surface albedo (α), that is, $K\uparrow = K\downarrow\alpha$.

The sky hemisphere is also the source of longwave radiation ($L\downarrow$), the magnitude of which is a function of both air temperature (T_a) and atmospheric emissivity (ε_a):

$$L\downarrow = f\left(\varepsilon_a, T_a^4\right) \tag{2.4}$$

The bulk of $L\downarrow$ at the ground originates in the denser lowest layer of the atmosphere where emissivity depends on the humidity of the air (ρ_v) and the amount and type of cloud cover. Cloud is an extremely efficient absorber and emitter of longwave radiation: the more extensive the cloud cover and the lower the cloud level, the higher the atmospheric emissivity. Typically, ε_a varies between 0.7 (low humidity and clear skies) and >0.9 (extensive low cloud cover). The magnitude of outgoing longwave radiation ($L\uparrow$) is a function of the surface temperature (T_s) and emissivity (ε_s):

$$L\uparrow = f\left(\varepsilon_s, T_s^4\right) \tag{2.5}$$

The emissivity of most natural and manufactured materials is between 0.85 and 0.95 (Table 2.1), such that the reflection of $L\downarrow$, which equals $(1 - \varepsilon_s)$, is small.

On a clear day the major driver of diurnal radiation exchanges is $K\downarrow$, which is zero at night and peaks at noon; $K\uparrow$ is a fraction of $K\downarrow$ and varies in direct response. The longwave radiation exchanges are relatively consistent by day and night but $L\uparrow$ is appreciably larger than $L\downarrow$, such that $L\downarrow$-$L\uparrow$ is typically about $-100\,\mathrm{W\,m^{-2}}$. Net radiation (Q^*) is symmetrical around noon but is positive during the daytime and negative at nighttime. In other words, the ground warms by radiation between sunrise and sunset and cools by radiation at night (Fig. 2.4). The magnitude of these diurnal patterns will change with latitude and the seasons that control the length of daylight and the intensity of $K\downarrow$. Moreover, the diurnal radiation-exchange patterns are rarely smooth, as $K\downarrow$ is interrupted by the passage of clouds. If cloud cover is low and extensive, the magnitude of $K\downarrow$ will be very low and there is little energy available for ground-air (and ground-soil) exchanges. However, just as cloud cover reduces solar gain during the daytime, it also restricts longwave radiation loss at night.

Net radiation (Q^*) can be regarded as available energy at the ground that can be partitioned into sensible (Q_H) and latent (Q_E) heat exchange by convection with the overlying atmosphere and sensible heat exchange with the substrate by conduction (Q_G):

$$Q^* = Q_H + Q_E + Q_G \tag{2.6}$$

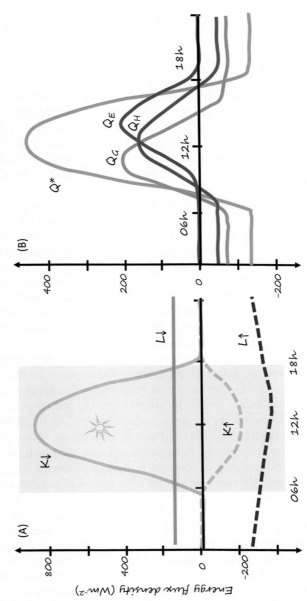

Fig. 2.4 A general depiction of the (A) surface radiation budget and (B) surface energy budget over the course of a sunny day for a grassland surface.

Each of the nonradiative exchanges are functions of a vertical gradient (Δ) of a relevant property and a means of exchange:

$$Q_G = f\left(\Delta T_{sub}, k\right) \tag{2.6a}$$

$$Q_H = f\left(\Delta T_a, K_H\right) \tag{2.6b}$$

$$Q_E = f\left(\Delta \rho_v, K_V\right) \tag{2.6c}$$

In the case of Q_G, the flux is regulated by the gradient in substrate temperature (ΔT_{sub}, $K\,m^{-1}$) and the conductivity (k) of the soil material. The convective fluxes, Q_H and Q_E, are regulated by the gradients in air temperature, ΔT_a ($K\,m^{-1}$), and water vapor, $\Delta \rho_v$ ($kg\,m^{-1}$), and the respective eddy conductivities, K_H and K_V. The eddy conductivity terms are shorthand expressions for complex convective processes that mix the atmosphere and account for the effect of both wind and atmospheric stability.

Fig. 2.4 presents the typical diurnal patterns for the radiative (Fig. 2.4A) and nonradiative (Fig. 2.4B) fluxes over a grassland surface under clear and calm conditions. At night, the ground cools by radiation ($Q^* < 0$) and this loss is compensated for by the transfer of energy from the substrate ($Q_G < 0$) and from the overlying air toward the surface ($(Q_H + Q_E) < 0$). Similarly, during the daytime the ground is heated by radiation and energy is transferred to the atmosphere and the substrate, warming the substrate and near-surface air and increasing its vapor content. Note that the diurnal patterns of exchanges are not symmetrical as they depend on the relevant gradients (ΔT_{sub}, ΔT_a, and $\Delta \rho_v$) and transfer terms (k, K_H, and K_V), each of which will change in response to the respective energy exchanges (Q_G, Q_H, and Q_E).

After sunrise, Q^* increases and becomes positive. Initially, this energy is preferentially channeled into the substrate (Q_G), while the response of the convective exchanges (Q_H and Q_E) lags. Q_G reaches a peak before noon and declines in the afternoon, becoming negative before sunset. By comparison, Q_H and Q_E peak after noon before declining and becoming zero or negative after sunset. The offset between the conductive and convective fluxes can be attributed to the behavior of the substrate and near-surface atmosphere, respectively. In the morning hours, the substrate is cool and ΔT_{sub} is large; by comparison, the atmosphere is stable so although ΔT_a ($\Delta \rho_v$) may be large, K_H (K_V) is weak. By noon, the substrate has been warmed and, although the surface continues to warm, ΔT_{sub} has been reduced and Q_G is lowered. As the surface warms, it heats the overlying air, which becomes progressively less stable (more turbulent) making mixing easier. The turbulent fluxes (Q_H and Q_E) increase during the late morning and early afternoon. As the surface cools in late afternoon, the intensity of turbulence weakens, the fluxes diminish and become negative after sunset. Note that in Fig. 2.4, $Q_E > Q_H$ for much of the daytime indicating that for a surface where water is readily available, Q^* is preferentially channeled into the evaporation of water through plants (referred to as evapotranspiration). A useful measure of the relative roles of these two fluxes is the Bowen ratio (β), which is simply the ratio Q_H/Q_E; if $\beta < 1$, evaporation dominates and if $\beta > 1$, heating dominates.

Practically, we do not measure fluxes at the ground but at a height above the surface. While the energy terms must balance at the ground surface, to consider a volume (in which energy can be stored) we need to examine its energy budget. Fig. 2.3 displays a control volume that includes the near-surface air and underlying substrate. The lower plane is placed at a level in the substrate where the diurnal temperature range is zero, so that here Q_G equals zero:

$$Q^* = Q_H + Q_E + \Delta Q_A + \Delta Q_S \tag{2.7a}$$

One of the new terms (ΔQ_A) accounts for the advection of sensible and latent heat flux across the sides of the control volume, but if the site is over an extensive horizontal grass surface, then $\Delta Q_A = 0$. The remaining term (ΔQ_S) represents the storage of energy in this volume as a result of net energy gain and loss:

$$\Delta Q_S = Q^* - (Q_H + Q_E) \tag{2.7b}$$

In words, then, measuring Q^*, Q_H, and Q_E at the upper plane of the volume should be sufficient to "close" the budget. As there is little heat stored in the atmospheric part of the volume, the bulk of ΔQ_S is contained in the substrate, which represents the thermal memory in the system. A useful measure of the capacity of the substrate to store sensible heat is given by thermal admittance (or thermal inertia):

$$\mu = \sqrt{kC} \tag{2.8}$$

Thermal admittance combines two thermal properties of the substrate: conductivity (k), which is the ability to transfer sensible heat along a gradient; and heat capacity (C), the thermal response of the material to heat added/lost (Table 2.2). The lower (higher) the value for admittance, the faster (slower) the substrate can store and release heat. Fig. 2.4 shows that during the early morning, energy is being stored in the substrate, but from late evening onwards, this energy is removed. It is important to note that the thermal properties of soil can change drastically, depending mainly on its water/air content; the drier (wetter) soil becomes, the lower (higher) the admittance value.

2.2.1.1 The temperature response

The energy fluxes cause changes to the temperature of the substrate and to the adjacent air, which modifies ΔT_{sub} and ΔT_a and alters the fluxes (Fig. 2.5). During the morning, the convergence of Q_G, in a series of layers at increasing depth, causes the profile to change as a "wave" of heating extends downwards. By late afternoon the ground starts to cool, the ΔT_{sub} near the ground reverses (that is, the soil is warmer), even as heat may be transferred downwards at deeper layers. Q_G is now directed toward the ground surface and this sustains T_s. The cooling process draws on the heat stored in the morning and the substrate temperature profile responds. By comparison with wet soil, the reservoir of heat in dry soil is small and is quickly depleted, and just as the ground surface warmed quickly during the daytime, it cools quickly at night.

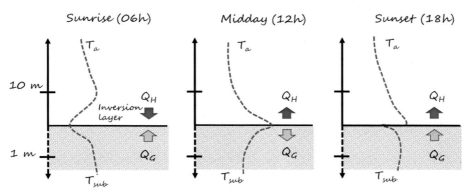

Fig. 2.5 The thermal response of the near-surface atmosphere and substrate to the energy exchanges depicted in Fig. 2.4 at sunrise, midday, and sunset. The ground cools overnight due to longwave radiation loss and heat is removed from the overlying air ($Q_H < 0$) and substrate ($Q_G < 0$), generating inverted temperature profiles. By midday, the ground has been warmed by the Sun and heat is transferred to the air ($Q_H > 0$) and substrate ($Q_G > 0$). By sunset, the substrate has started to cool ($Q_G < 0$) but the surface continues to warm the overlying air ($Q_H > 0$).

A similar process happens in the overlying atmosphere except that the eddy conductivity (K_H) also changes in response to wind speed (v), the roughness of the surface (z_0), and atmospheric stability (φ):

$$K_H = f\left(v, z_0, \phi\right) \tag{2.9}$$

The cause of mixing can be attributed to forced or free convection, or a combination of both. Mixing by forced convection occurs as airflow is slowed by friction with the Earth's surface—the stronger the wind speed and the rougher the surface, the greater is the forced mixing. Free convection is linked to the thermal stratification of the atmosphere. Overnight, cooling of the near-surface layer of air produces a temperature inversion ($\Delta T_a < 0$) that impedes convection. After sunrise, surface warming eventually heats this layer sufficiently to generate convection. Thereafter, mixing extends the influence of the warming surface upwards as warmer air closer to the ground is displaced and cooler air is drawn downwards. The magnitude of ΔT_a governs the intensity of mixing and, once ΔT_a exceeds $0.01\,°C\,m^{-1}$, the atmosphere is unstable and vertical mixing occurs readily. After sunset, the ground begins to cool, Q_H reverses, and the near-surface air cools as the surface extracts heat from the adjacent air.

2.2.2 Paved surfaces

Construction materials are generally designed to be strong, impermeable, and resilient. Paving imposes a seal over the underlying soil substrate, preventing the exchange of water and gases with the overlying air. The absence of the latent heat flux means that the surface energy balance is simplified:

$$Q^* = Q_H + Q_G \tag{2.10}$$

A large, flat asphalt surface is geometrically nearly identical to the grass-covered surface discussed above and presents little roughness to air motion. Under identical conditions, K↓ values for the asphalt and the grass surfaces are the same, but the lower albedo for asphalt means that K↑ will be lower (Table 2.1). L↓ on this surface may be slightly higher than that over a natural surface, as the overlying air is likely to warmer (Eq. 2.4). However, the absence of Q_E means that all available energy is expended as sensible heat fluxes (Q_H and Q_G). As a result, T_s is raised and so is L↑. Overall, the changes to the radiation terms compensate so that Q* is close in magnitude to that discussed for the natural surface.

Fig. 2.6 illustrates the importance of available water on T_s for a typical summer's day in an arid climate (Phoenix, USA). Measurements were made over three urban sites: a suburban site with irrigated grass; an urban site over desert (the natural landscape); and an urban site over asphalt. Surface temperature was measured using a downward-facing infrared thermometer mounted at a height of 3 m, observing a circular area of $0.5\,m^2$ on the ground. The diurnal temperature curves for each site reveal dramatic differences in T_s of up to 30 °C between the grass and desert/asphalt sites during the daytime. The grass surface can use available energy to evaporate water into the overlying air, and the presence of water in the underlying soil increases its thermal

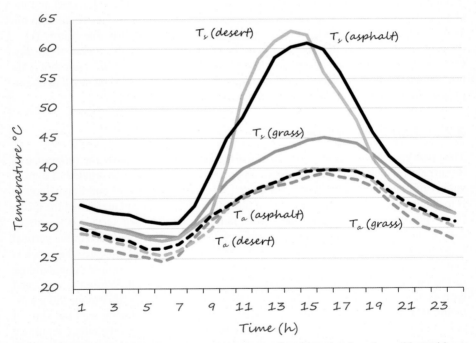

Fig. 2.6 Surface (T_s) and air (T_a) temperatures measured at three sites in a dry and hot arid environment (Phoenix, Arizona) in late summer. The sites represent asphalt, watered grass, and the natural cover (desert).

Source: Modified after Stoll, M,J., Brazel, A.J., 1992. Surface-air temperature relationships in the urban environment of Phoenix, Arizona, Phys. Geogr. 13, 160–179.

admittance. The net effect is that the diurnal variation in T_s over grass is moderate. By comparison, the desert and asphalt surfaces experience similar daytime highs in T_s as neither surface has water to evaporate. Note that the difference between T_s for each surface type depends on which is selected as the "natural" one. If the desert is selected, then T_s over grass is cooler but if the grass surface is selected, then both the desert and asphalt surfaces are much hotter during the daytime.

Fig. 2.6 also shows measured air temperature over each surface type. While the response of T_a over the grass surface matches that of T_s, the daytime T_a over asphalt and desert does not follow the underlying T_s. This discrepancy can be attributed to the scale of the drivers that affect surface and air temperatures. While the grass surface was extensive, the urban sites were situated in a more heterogeneous area. As a result, the air over the grass remained closely coupled to the underlying surface, but the air over the other surfaces was constantly replaced with "new" air as a result of advection.

More generally, the difference between the T_s of grass and asphalt depends primarily on the state of the natural surface. Critically, the thermal properties of the manufactured materials are consistent, as their water content does not change significantly. By comparison, the properties of the soil change dramatically with water content, which raises both conductivity (k) and heat capacity (C). This has the effect of making the soil a good store for sensible heat and depressing the surface (and overlying air) temperature (Table 2.2). As soil dries, the values of k and C are lowered and the change from wet to dry can switch the position of the natural surface relative to the asphalt surface, so that, despite receiving the same amount of available energy (Q^*), the asphalt may be warmer or cooler than the grass surface over the course of the day. Other properties of manufactured materials (such as the albedo and emissivity) do not change with the seasons, unlike natural surfaces. It is this consistency that distinguishes most manufactured materials from natural surfaces.

2.2.3   Building surfaces

The outer envelope of buildings is distinguished from flat ground by the properties of its surfaces, which have distinct slopes and orientations and are constructed of a great variety of materials (stone, wood, glass, etc.). The primary effect of building geometry is to change the timing and magnitude of the solar radiation receipt on each of the facets (wall and roof). For a simple cube-shaped building, the flat roof surface is geometrically identical to the asphalt and grass surfaces discussed above, but each of the vertical surfaces experiences distinct diurnal solar radiation patterns that indicate when each is "viewed" by the Sun. Fig. 2.7 demonstrates these patterns for a simple cube building in the Northern Hemisphere oriented with a south-facing wall. While the east- and west-facing walls receive peak solar radiation in morning and afternoon, the south-facing wall receives peak radiation at noon, as does the north-facing wall but since it is in shade throughout the day, this is diffuse solar radiation only. The effect of surface geometry will be readily apparent in the relative temperatures of these facets, which will experience different patterns of daytime heating and cooling. The diffuse radiation terms also differ by facet, as each of the walls will receive some solar radiation via reflection from the adjacent ground, while the roof receives radiation from the sky only.

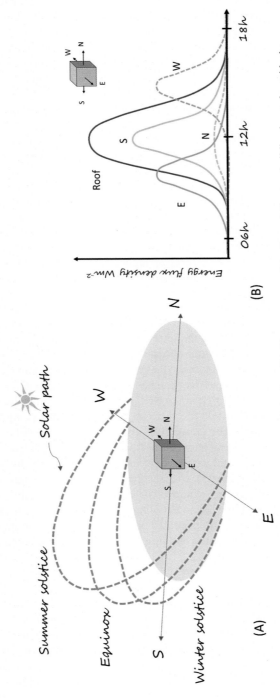

Fig. 2.7 The relation between a cube-shaped building and the Sun. (A) The movement of the Sun relative to a building located outside the tropics, and (B) the pattern of energy interception on each of the cube facets.

Longwave radiation receipt (L↓) also varies by building facet owing to the different sky view factors of the roof (SVF=1) and each of the walls (SVF=0.5). This means that while the roof gains L↓ from the sky, the walls receive longwave radiation from both the sky and ground. As the ground will usually have a higher emissivity than the atmosphere, the walls will have a higher receipt of L↓. Keep in mind that view factors are reciprocal and that the SVF of the ground near the walls will be restricted. As a consequence, the walls and adjacent ground will "recycle" much of the longwave radiation as L↑ emitted by either surface will be received by the other (as L↓), and vice versa. By comparison, little of the radiation emitted to the sky is returned because the sky is normally a radiative "sink." The impact of the facet differences in longwave radiation exchanges is clearest overnight in clear and calm conditions when convective transfer (Q_H) is weak and surfaces cool mainly through longwave radiative heat loss (L↑ - L↓).

While the surfaces will have different T_s values depending on when they experienced daytime heating, they will also cool at different rates. Comparing the east- and west-facing walls, the latter received most shortwave radiation in the afternoon when direct solar receipt was enhanced by radiation reflected from the adjacent, sunlit ground. In addition, this wall also receives higher L↓ from this warmed ground. At night, the roof cools quickest as it has the highest SVF and experiences the greatest net loss of longwave radiation.

Net radiation (Q*) at the building envelope is partitioned into Q_H and Q_G. The conductive flux is controlled by the difference between the surface and interior temperatures and the thermal properties of the intervening fabric(s). Since a primary purpose of the building is to create an indoor climate suited to its inhabitants, the role of the fabric is to manage heat gain and loss via the envelope. Internal heating and/or cooling systems (both mechanical and passive) can be used to address excessive heat loss and gain, respectively. As a result, Q_G is controlled by both the external natural drivers and internal anthropogenic drivers and the flux can be negative (that is, toward the outer surface) during the daytime. The convective process at the walls and roof is complicated by the fact that the building obstructs airflow, generating turbulence and creating gusts and lulls that enhance and depress heat exchanges, respectively. Moreover, the envelope is rarely sealed completely, so that air from the outdoors enters and indoor air escapes either by design (opening windows) or through gaps in the fabric (infiltration). Finally, the exhausts of building heating/cooling systems expel waste heat directly into the outdoors. These outlets are often located on walls or are concentrated on roofs.

The overall energy budget for the building volume then can be stated as:

$$Q^* + Q_F = Q_H + \Delta Q_S \tag{2.11}$$

The average Q^* for the building envelope will look similar to that of the paved flat surface with a peak at noon and symmetrical morning and afternoon limbs. The new term, Q_F, represents the anthropogenic heat flux, i.e., the energy generated internally. Q_F varies over the course of the day/week in response to occupation patterns and the outdoor climate, which drives the need for heating/cooling. The storage term (ΔQ_S) is concentrated in the building fabric and in the ground below the building.

2.2.3.1 Outdoor microclimatic impacts

For the flat surfaces described previously, their impact on adjacent surface types is indirect, via advection. By comparison, the impact of large 3D features, such as buildings, on the adjacent ground includes both direct (overshadowing) and indirect effects on radiation exchanges and atmospheric mixing. The length of shadow cast is a function of the height of the object and the Sun's zenith angle. Between 0 and 60°, the shadow length increases approximately linearly: at 45°, it is the same height as the object (H); and at 63 ½ °, it is twice the height (2H). Thereafter it increases quickly and at 80° the length is 5.7H (Fig. 2.8). A building therefore creates a shaded area in the opposite direction to the Sun's azimuth based on its width and height. This area represents the direct radiation intercepted by the building and the corresponding loss of direct radiation to the shaded ground. Not surprisingly, buildings of modest height can have dramatic effects on the solar climatology of the adjacent ground at high latitudes or in winter at mid-latitudes (Fig. 2.8) (Atkinson, 1912).

Buildings are solid features that disturb the ambient airflow by blocking its passage and displacing the air around its sides and over the roof. To the rear of the building, a series of eddies (rotating vortices) are generated of varying sizes that are in place for a

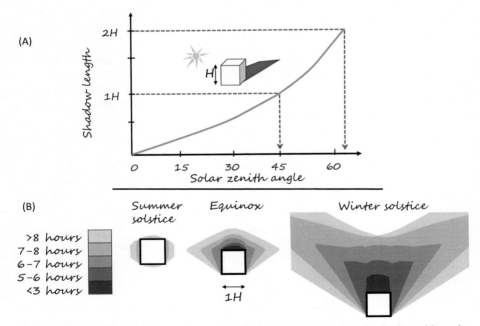

Fig. 2.8 (A) The length of shadow for a cube-shaped building based on the Sun's zenith angle. (B) The pattern of shadow and hours of direct sunlight available around a building at 40° latitude.

Adapted from Atkinson, W., 1912. The Orientation of Buildings: Or, Planning for Sunlight. J. Wiley & Sons, New York. Available at https://archive.org/details/orientationofbui00atki/page/n5/mode/2up.

period of time before being advected downwind. The exact pattern of disturbance depends on the shape of the obstacle and the steadiness of the ambient airflow (speed and direction). As the wind increases rapidly with height, taller buildings have a greater impact than lower buildings by drawing faster winds to the ground. The net effect of buildings is to increase atmospheric mixing by forced convection in their vicinity, enhancing turbulent exchanges. The net impact of isolated buildings is to dramatically alter the nearby surface climates.

2.3   The urban landscape

Cities are comprised of a myriad of surface types (both natural and manufactured), many of which are small in extent and have complex geometries. This spatial heterogeneity means that considerable microclimatic variability occurs over very short distances. While the temperature at the surface will bear the distinct imprint of its thermal and radiative properties, that of the near-surface atmosphere will show the influences of many different surface types that have been acquired over a wider area. One of the challenges that faces the UHI researcher is how best to describe (and organize) this heterogeneity in a meaningful way.

Oke et al. (2017) classified the urban landscape into a hierarchy of scales, each of which exerts a distinct effect on the adjacent atmosphere. In this system, the city can be decomposed into progressively smaller features: neighborhoods, blocks, streets, buildings, and facets (Fig. 2.9). A *facet* is a planar surface of homogenous fabric and orientation (such as paved ground; Fig. 2.10). A *building* is enclosed by facets that constitute the envelope; each facet will have a different material composition and/or orientation. The *street* is the first, recognizably, urban scale as it represents the geometry of the outdoor spaces formed between buildings (Fig. 2.11). The *block* is formed by intersecting streets that enclose several buildings (Fig. 2.12). The diverse orientations and aspects of streets results in different rates of daytime heating and nighttime cooling, even if all streets have the same material fabric. The *neighborhood* describes an urban landscape of $> 1\,km^2$, which includes a diversity of facets and 3D elements. However, each neighborhood type has a characteristic combination of features that distinguishes it from other types, often associated with the typical functions present (such as residential or warehouse storage). Finally, the *city* is the entire urbanized landscape, which is made up of neighborhoods of varying types and extents.

2.3.1   City streets

Much of what we know of the urban effects on temperature is based upon work done on simple city streets (known as urban canyons). A canyon is a long street that is symmetric in profile and has little or no vegetation (Fig. 2.11). Its geometry consists of two parameters: the aspect ratio, or height-to-width (H/W) ratio, and orientation (φ) of the street axis. The aspect ratio is an especially important descriptor as it links several of the climatic impacts (solar access, sky view factor, and wind) associated with street design. H/W and φ affect the timing and distribution of energy exchanges

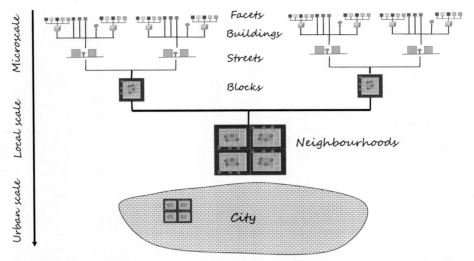

Fig. 2.9 The urban landscape can be decomposed into elements organized by scale and structure. Each element has a unique climatic effect that depends on its individual properties and it relation with neighboring elements. Facets are planar elements made of consistent materials (e.g., glass, asphalt, turf) with associated thermal and radiative properties that can be assembled to make streets, gardens, and building envelopes. The dimensions and placements of buildings and trees along streets creates the three-dimensional geometry of the urban canopy layer. The layout of groups of buildings and open spaces creates blocks and neighborhoods at the local scale, with distinctive properties (e.g., average building heights, fraction of impervious surface cover). The complete city footprint consists of neighborhood types that are draped over the underlying terrain.

Fig. 2.10 A variety of urban facets made of manufactured and natural materials. From top and left to right: a modern building wall in Sao Paulo (Brazil); a rooftop in Naples (Italy); corrugated iron roofs in Soweto (South Africa); ground covered by paving and grass in Dublin (Ireland); stone paving in Parati (Brazil); and a stone wall in Wexford (Ireland). From G. Mills, except for Soweto image from Matt-80 and licensed under Creative Commons Attribution-Share Alike 2.0 Generic.

Fig. 2.11 Street landscapes. The two images on the left are from Dublin (Ireland) and show compact midrise (top) and open lowrise (bottom) neighborhoods; the image on the right shows a compact highrise neighborhood in Hong Kong (China).
From G. Mills.

Fig. 2.12 Blocks and neighborhoods viewed from above. From the top and left to right: Sparks (USA); Dublin (Ireland); London (UK); Lisbon (Portugal); Bangkok (Thailand); and Sao Paulo (Brazil).
From G. Mills, except for Sparks warehouse district by K. Lund, licensed under the Creative Commons Attribution-Share Alike 2.0 Generic.

within the street, and between the street and the overlying atmosphere. From an energy balance perspective, the exchanges at each of the canyon facets (walls and street surfaces) can be evaluated but so too can the exchanges across the top of the street. The energy budget of the canyon volume accounts for the exchanges at this level and changes to heat storage (ΔQ_S) in the bounding walls and street; keep in mind that walls are the shared interface with the indoor building spaces, which are sources of anthropogenic heat.

Street geometry has several distinct impacts on energy exchanges within the canopy (Fig. 2.13). First, H/W and φ together control direct solar receipt at canyon surfaces by presenting surfaces at different angles to the solar beam and creating shadow patterns (Fig. 2.12A). Second, H/W is linked to the SVF and thus regulates the origin and destination of diffuse radiation exchanges within the street: the smaller the SVF, the greater the opportunity for recycling radiation among the canyon facets. This has two consequences:

1. Shortwave radiation is reflected multiple times within the canyon before exiting (Fig. 2.13B). While the albedo of each facet is fixed, the net effect is to lower the albedo of the canyon unit, making it a relatively efficient absorber of K↓.
2. Longwave radiation is emitted from one facet to another, also known as the horizon screening effect (Fig. 2.13C). As the sky is normally a longwave radiation sink, reducing the SVF limits net longwave (L↓ - L↑) loss and slows the radiative cooling process at facets. While the emissivity of each facet is fixed, the net effect is to increase the emissivity of the canyon unit making it a more efficient absorber (and emitter) of longwave radiation.

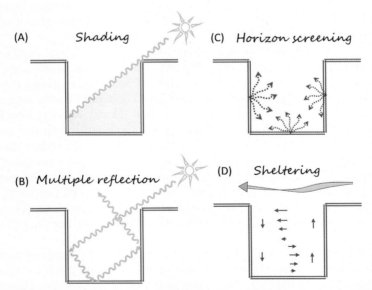

Fig. 2.13 The impact of street geometry on (A) direct and (B) diffuse shortwave radiation and (C) diffuse longwave radiation exchanges within the urban canopy layer. Street geometry also impacts on (D) convective exchange by providing shelter.

Finally, street geometry alters near-surface airflow, usually providing a sheltered environment and dampening the turbulent heat exchanges at each facet (Fig. 2.13D). The relation between canyon airflow and ambient flow (above roof level) is complex and depends on H/W and the ambient wind direction relative to street orientation. The sheltering effect is strongest when the street is narrow (H/W < 1) and ambient airflow is perpendicular to the canyon axis.

There have been very few measurements of the surface energy balances within urban canyons. Fig. 2.14A and B shows the measured net radiation (Q^*) fluxes at the walls of a north-south oriented alley in Vancouver (Canada) over a summer (Nunez and Oke, 1977). Note the differences between the sides of the canyon, with one side receiving the bulk of its solar input in the morning, and the other side in the afternoon. The effects of multiple reflection are seen in the secondary peaks on both walls that correspond with the maximum on the opposite wall. Note also that when Q^* for the canyon unit is obtained (that is, the sum of Q^* at the walls and floor divided by the area at roof height), the diurnal pattern is very similar to that over the paved flat surface discussed previously (Fig. 2.14C).

2.3.2 Neighborhoods

Neighborhoods consist of common arrangements of buildings, roads, parks, paving, etc., that are often associated with particular functions (such as residential, commercial, and industrial). A neighborhood is a local-scale landscape feature that covers an area of over $1 \, km^2$ and encompasses several types of microclimates. To evaluate the energy fluxes that characterize a neighborhood type, instruments must be positioned (i) well above the ground and building cover so that the microclimatic contributions are blended, and (ii) far from the windward edge of that neighborhood to remove any upwind effects (that is, $\Delta Q_A = 0$).

Fig. 2.15 shows an observation platform for measuring Q^*, Q_H, and Q_E over a mid-latitude city-center neighborhood. The instruments are placed above the roughness sublayer (Fig. 1.2) at over twice the height of buildings, capturing the signals from a diverse urban landscape below. Net radiation is obtained from the measurement of shortwave (K) and longwave (L) radiation toward ($\downarrow$) and away ($\uparrow$) from the surface using pyranometers (K) and pyrgeometers (L). The turbulent sensible and latent heat terms are obtained using the eddy correlation method, which obtains these from joint correspondence between short-term fluctuations in vertical wind speed and of air temperature (T_a) and vapor density (ρ_v), respectively. One of the challenges of site selection is matching the fixed area from which the radiation (K$\uparrow$ and L$\uparrow$) signals are received with the fluctuating areas from which the turbulent terms (Q_H and Q_E) are received, so that they represent the same neighborhood type and Eq. (2.7) can be applied.

Fig. 2.14D shows the energy budget of a densely built neighborhood in Mexico City with little vegetation. The major feature of this neighborhood is the absence of any substantial evaporation ($Q_E \approx 0$), so that nearly all of the available energy is expended as sensible heat exchange with the overlying atmosphere (Q_H) or is stored in the substrate (ΔQ_S). It is important to distinguish between observations made above

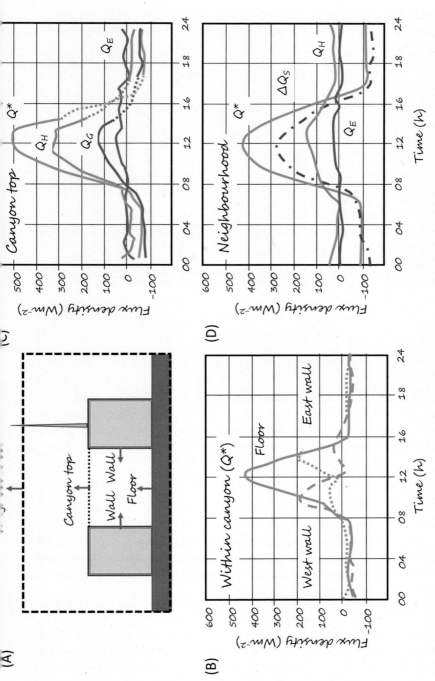

Fig. 2.14 Energy flux observations at (A) three levels for (B and C) a street canyon in a commercial area of Vancouver (Canada) during July, and (D) a residential neighborhood in Mexico City during December. The Vancouver site is a north-south oriented alley where net radiation was measured at (B) the wall and floor facets. The energy balance at (C) the canyon top represents the aggregate contribution of the facets. The energy fluxes in (D) were observed over a compact neighborhood during six almost cloudless days.

Redrawn from Nunez, M., Oke, T.R., 1977. Energy balance of an urban canyon. J. Appl. Meteorol. 16, 11–19, and Oke, T.R., Spronken-Smith, R.A., Jáuregui, E., Grimmond, C.S.B., 1999. The energy balance of Central Mexico City during the dry season. Atmos. Environ. 33, 3919–3930.

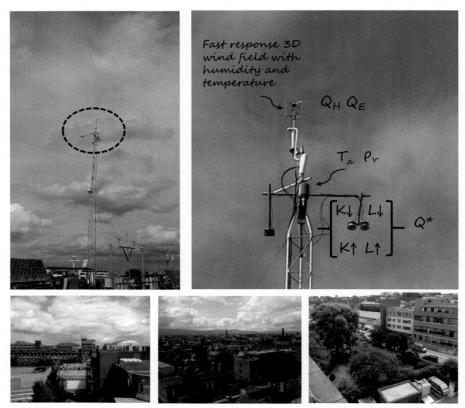

Fig. 2.15 A flux tower located over a compact midrise neighborhood in Dublin (Ireland). The top row shows the instrumented tower positioned above a rooftop, and the instruments used to measure the incoming ($\downarrow$) and outgoing ($\uparrow$) shortwave (K) and longwave (L) radiation terms; the air temperature (T_a) and humidity (ρ_v); and the sensible (Q_H) and latent (Q_E) heat fluxes. The bottom row shows the surrounding urban landscape from which the tower acquires its signals.
From G. Mills and S. Keogh.

the urban canopy layer (UCL) and those made within (Fig. 2.14B, C). The former integrate the microclimatic contributions of the UCL, which have been processed and exchanged at the top of the open canopy, *plus* those at and above roof level below the measurement plane (Fig. 2.14).

Climatologists have employed a variety of land-use/land-cover (LULC) classifications to impose a spatial structure on the urbanized landscape. These can be used as a framework to design an urban observation program and analyze its output. Ideally, such a framework should be based on an understanding of surface-air energy exchanges, such that its classes represent climatically relevant (and recognizable) features of cities. The Local Climate Zone (LCZ) classification specifies urban and rural landscapes into 17 classes, of which 7 are land-cover types (LCZs A–G) and

10 are built types (LCZs 1–10)—see Table 2.3 (Stewart and Oke, 2012). Each LCZ class is distinguished by the heights and proximity of three-dimensional roughness elements; the fractional cover of paved and building surfaces; and by characteristic values of albedo, surface thermal admittance, sky view factor, and anthropogenic heat flux (Table 2.4). For example, LCZ 1 describes a compact highrise area typical of a central business district that has tall, closely spaced buildings, little or no vegetation, and a large H/W ratio (small SVF). LCZ 6, in contrast, describes an open low-rise

Table 2.3 Definitions for Local Climate Zones.

Built types	Definition	Land cover types	Definition
1. Compact highrise	Dense mix of tall buildings to tens of stories. Few or no trees. Land cover mostly paved. Concrete, steel, stone, and glass construction materials.	A. Dense trees	Heavily wooded landscape of deciduous and/or evergreen trees. Land cover mostly pervious (low plants). Zone function is natural forest, tree cultivation, or urban park.
2. Compact midrise	Dense mix of midrise buildings (3–9 stories). Few or no trees. Land cover mostly paved. Stone, brick, tile, and concrete construction materials.	B. Scattered trees	Lightly wooded landscape of deciduous and/or evergreen trees. Land cover mostly pervious (low plants). Zone function is natural forest, tree cultivation, or urban park.
3. Compact lowrise	Dense mix of lowrise buildings (1–3 stories). Few or no trees. Land cover mostly paved. Stone, brick, tile, and concrete construction materials.	C. Bush, scrub	Open arrangement of bushes, shrubs, and short, woody trees. Land cover mostly pervious (bare soil or sand). Zone function is natural scrubland or agriculture.
4. Open highrise	Open arrangement of tall buildings to tens of stories. Abundance of pervious land cover (low plants, scattered trees). Concrete, steel, stone, and glass construction materials.	D. Low plants	Featureless landscape of grass or herbaceous plant cover. Few or no trees. Zone function is natural grassland, agriculture, or urban park.
5. Open midrise	Open arrangement of midrise buildings (3–9 stories). Abundance of pervious land cover (low plants, scattered trees). Concrete, steel, stone, and glass construction materials.	E. Bare rock or paved	Featureless landscape of rock or paved cover. Few or no trees or plants. Zone function is natural desert (rock) or urban transportation.
6. Open lowrise	Open arrangement of lowrise buildings (1–3 stories). Abundance of pervious land cover (low plants, scattered trees). Wood, brick, stone, tile, and concrete construction materials.	F. Bare soil or sand	Featureless landscape of soil or sand cover. Few or no trees or plants. Zone function is natural desert or agriculture.
7. Lightweight lowrise	Dense mix of single-story buildings. Few or no trees. Land cover mostly hard-packed. Lightweight construction materials (e.g., wood, thatch, corrugated metal).	G. Water	Large, open water bodies such as seas and lakes, or small bodies such as rivers, reservoirs, and lagoons.
8. Large lowrise	Open arrangement of large lowrise buildings (1–3 stories). Few or no trees. Land cover mostly paved. Steel, concrete, metal, and stone construction materials.	**VARIABLE LAND COVER PROPERTIES**	
9. Sparsely built	Sparse arrangement of small or medium-sized buildings in a natural setting. Abundance of pervious land cover (low plants, scattered trees).	Variable or ephemeral land cover properties that change significantly with synoptic weather patterns, agricultural practices, and/or seasonal cycles.	
		b. bare trees	Leafless deciduous trees (e.g., winter). Increased sky view factor. Reduced albedo.
		s. snow cover	Snow cover > 10 cm in depth. Low admittance. High albedo.
10. Heavy industry	Lowrise and midrise industrial structures (towers, tanks, stacks). Few or no trees. Land cover mostly paved or hard-packed. Metal, steel, and concrete construction materials.	*d. dry ground*	Parched soil. Low admittance. Large Bowen ratio. Increased albedo.
		w. wet ground	Waterlogged soil. High admittance. Small Bowen ratio. Reduced albedo.

Source: Stewart, I.D., Oke, T.R., 2012. Local Climate Zones for urban temperature studies. Bull. Am. Meteorol. Soc. 93, 1879–1900. Reproduced with permission from the American Meteorological Society.

Table 2.4 Parameter values for Local Climate Zones.

Local Climate Zone	Height of roughness elements [a]	Building surface fraction [b]	Impervious surface fraction [c]	Pervious surface fraction [d]	Sky view factor	Anthropogenic heat flux, Q_F (Wm^{-2})
1. Compact highrise	> 25	40–60 %	40–60 %	< 10 %	0.2–0.4	50–300
2. Compact midrise	10–25	40–70 %	30–50 %	< 20 %	0.3–0.6	< 75
3. Compact lowrise	3–10	40–70 %	20–50 %	< 30 %	0.2–0.6	< 75
4. Open highrise	> 25	20–40 %	30–40 %	30–40 %	0.5–0.7	< 50
5. Open midrise	10–25	20–40 %	30–50 %	20–40 %	0.5–0.8	< 25
6. Open lowrise	3–10	20–40 %	20–50 %	30–60 %	0.6–0.9	< 25
7. Lightweight lowrise	2–4	60–90 %	< 20 %	< 30 %	0.2–0.5	< 35
8. Large lowrise	3–10	30–50 %	40–50 %	< 20 %	> 0.7	< 50
9. Sparsely built	3–10	10–20 %	< 20 %	60–80 %	> 0.8	< 10
10. Heavy industry	5–15	20–30 %	20–40 %	40–50 %	0.6–0.9	> 300
A. Dense trees	3–30	< 10 %	< 10 %	> 90 %	< 0.4	0
B. Scattered trees	3–15	< 10 %	< 10 %	> 90 %	0.5–0.8	0
C. Bush, scrub	< 2	< 10 %	< 10 %	> 90 %	0.7–0.9	0
D. Low plants	< 1	< 10 %	<10 %	> 90 %	> 0.9	0
E. Bare rock or paved	< 0.25	< 10 %	> 90 %	< 10 %	> 0.9	0
F. Bare soil or sand	< 0.25	< 10 %	< 10 %	> 90 %	> 0.9	0
G. Water	n/a	< 10 %	< 10 %	> 90 %	> 0.9	0

[a] Geometric average of building heights (LCZs 1–10) and tree/plant heights (LCZs A–F) (m)
[b] Ratio of building plan area to total plan area (%)
[c] Ratio of impervious plan area (paved, rock) to total plan area (%)
[d] Ratio of pervious plan area (bare soil, vegetation, water) to total plan area (%)
Source: Stewart, I.D., Oke, T.R., 2012. Local Climate Zones for urban temperature studies. Bull. Am. Meteorol. Soc. 93, 1879–1900.

area with abundant vegetation and a small H/W ratio (large SVF). LCZ D describes a landscape with low vegetation cover similar to the grassland setting for a standard meteorological station. The classes are not exclusive to geographic regions and can therefore be found in any city worldwide.

Meteorological observations of energy fluxes using the same approach as that for Mexico City (Fig. 2.14D) indicate that LCZ types have distinct energy flux characteristics. Fig. 2.16 displays the partitioning of the energy available for surface-air exchanges ($Q^* - \Delta Q_S$) into sensible (Q_H) and latent (Q_E) heat fluxes. This relation has been derived for mid-latitude summertime climates; note that a key variable that governs Q_H/Q_E is the fraction of the surface that is vegetated (λ_V). These data reveal that the relation is

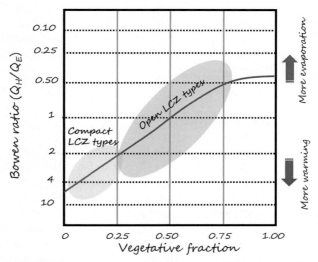

Fig. 2.16 Relation between the vegetative fraction of the surface and the partitioning of available energy into the convective fluxes, as represented by the Bowen ratio. The red line represents the average value measured over several weeks during summer at mid-latitude observation sites (above the UCL). The shaded areas indicate the ranges of values at compact and open LCZ urban sites (Table 2.4).
Based on Figure 6.17 in Oke, T.R., Mills, G., Christen, A., Voogt, J.A., 2017. Urban Climates. Cambridge University Press, Cambridge, UK.

approximately linear and that when $\lambda_V > 0.5$, the available energy is channeled into Q_E and conversely into Q_H when $\lambda_V < 0.5$. Clearly, to boost Q_E (evaporation) at the expense of Q_H (warming), increasing vegetative cover is an effective strategy.

2.3.3 Modelling the urban energy balance

There are a number of urban energy balance models that can simulate surface-air exchanges in response to land cover, surface geometry, and meteorological conditions. These vary in their sophistication with respect to the atmospheric processes and the description of the urban surface. One of these models is the Surface Urban Energy and Water Balance Scheme (SUEWS), which simulates radiation, energy, and water balances using only standard meteorological variables (air temperature, wind, and relative humidity) and information about the surface cover. Some of this information can be derived from the LCZ classification (Tables 2.3 and 2.4). SUEWS is designed for neighborhood-scale simulations and generates the fluxes above the roughness sublayer (equivalent to the level of flux observations, Fig. 2.15). It describes the underlying urban surface in terms of seven surface types: paved, buildings, evergreen trees/shrubs, deciduous trees/shrubs, grass, bare soil, and water. The scheme cleverly uses the diurnal relation between heat storage (ΔQ_S) and net radiation (Q^*) to estimate the turbulent flux exchanges, along with well-established methods to simulate Q_E (Q_H is obtained as a residual). Typically, SUEWS is applied to a grid cell that comprises fractions of surface types, each of which exerts a distinct influence on the overlying atmosphere. It can also be applied to a domain that is divided into grid cells; however, it does not account for advection ($Q_A = 0$). The model has been evaluated and applied in a number of different circumstances and can generate T_s and near-surface T_a for use in land-use planning (Ward et al., 2016).

2.4 The urban heat island

The UHI is a response to the changes in surface geometry, land cover, and land use discussed in Section 2.2. These changes result in a surface UHI (SUHI) that affects the overlying air, forming the canopy-layer (CUHI) and boundary-layer (BUHI) heat islands. Of these three, the CUHI is most accessible and can be measured using readily available and inexpensive thermometers. Sampling the urban landscape to provide city-scale CUHI assessments is easily achieved by mounting instruments on mobile platforms. As a result, there is a very large body of published work on the CUHI in many cities around the world. The SUHI has received much less attention until recently, mainly because of the difficulty in making T_s measurements and sampling the diversity of urban landscapes. However, the availability of land surface temperature (LST) data acquired by thermal infrared (TIR) sensors that have been deployed on satellite platforms since the 1980s has added considerably to our knowledge. Although the energy exchanges that underpin the SUHI are the same as those that regulate the CUHI, linking these two phenomena is not straightforward.

2.4.1 The CUHI

The CUHI is evaluated using fixed and/or mobile platforms to sample T_a in an urban landscape, which is then compared with the temperature obtained at a site in the adjacent rural or "natural" landscape (ΔT_{u-r}). This is the Lowry (1977) model and its body of work has established that the CUHI:

- is largest at night during dry and calm weather conditions, under cloudless skies;
- grows in magnitude from the edge of the city toward the densely built part, where buildings are relatively tall and closely spaced (Fig. 1.4);
- grows in magnitude from sunset to 3–5 h after sunset and largely disappears by sunrise; it may even be reversed in the early morning warming phase.

Fig. 2.17 shows T_a measurements made at two sites (over urban and green surfaces) in Birmingham (U.K.) during a heatwave event. Note that during the daytime, the two temperatures match closely but toward evening as the surfaces cool they follow different paths: while the temperature over the grass surface falls rapidly, that over the urban surface cools slowly. The CUHI grows in magnitude for several hours as the natural surface cools quickly at first. After several hours, the natural surface begins to cool more slowly and gradually the temperature difference starts to diminish.

The fact that the CUHI is largest at night is a major clue to the driving processes, which are associated with cooling. During calm and clear weather, surface cooling is

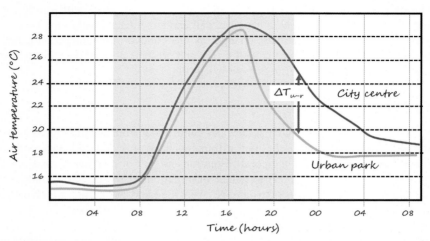

Fig. 2.17 The diurnal development of a canopy-level urban heat island (CUHI) on the night of 22 July 2013, during a heat wave in Birmingham (U.K.).
Redrawn from https://www.metlink.org.

dominated by longwave radiation heat loss as convective (and advective) exchanges are limited. A simplified energy budget under these circumstances states:

$$\left(L\downarrow - L\uparrow\right) \approx \Delta Q_s \tag{2.12}$$

The atmosphere is a poor absorber (and emitter) when there are no clouds, and little of the radiation emitted by the surface toward the sky is absorbed and re-emitted back to the surface. The weak winds mean that there is little mixing so the dominant nonradiative energy transfer is with the substrate. Over a grass surface with a large SVF, the ground will cool quickly, especially if the substrate is dry and heat stored during the day is quickly removed. By comparison, within the UCL the reduced SVF limits longwave radiation loss while the high thermal admittance (μ) of construction materials provides a reservoir of energy that can be withdrawn. Oke et al. (2017) suggested that in these conditions, the cooling differences between natural and urban near-surface air temperatures could be attributed to the divergence of longwave radiation in the layer of air next to the ground. The impact of the canopy geometry, then, is to slow radiative heat loss in the layer of air below roof height. A useful indicator of the maximum CUHI that can be expected under these ideal circumstances can be estimated from:

$$\Delta T_{u-r(\max)} = 15.27 - 13.88\,\text{SVF} \tag{2.13}$$

The sky view factor (SVF) is obtained in the center of the city where ΔT_{u-r} is strongest. Note that SVF is correlated with other urban parameters, such as low vegetative cover, the dominance of construction materials, and the addition of anthropogenic heat, each of which contributes to the formation of the UHI (Oke, 1981).

During the daytime, the CUHI is relatively small. In clear skies with strong solar radiation, temperature differences between wall and street facets within the UCL can be very large and exhibit dramatic changes as areas move into sunlight and shade. However, the canopy air volume is well mixed and presents none of the extreme temperature variations that the surfaces experience (Fig. 2.18).

2.4.2 The SUHI

The biggest contribution to SUHI studies has come about from satellite-based assessments of LST. Thermal infrared (TIR) instruments have provided LST data since the 1970s at varying spatial and temporal resolutions as part of a series of missions by Earth Observing Systems (EOS) to gather information on the oceans, clouds, land cover, and so on. Fig. 2.19 shows two images from the Landsat satellite, which gathers thermal information at a spatial resolution of 120 m at about 1000 h local time. The image pair shows the city and surrounding landscape of Atlanta (USA) on 28 September 2000. The urban landscape (roads, buildings, and pavement) appears on the visible image as grey, and on the thermal image the same pixels appear as warm (red). By comparison, the green areas appear as relatively cool (yellow). The difference in observed T_s between the warmest and coolest pixels of the thermal image is close to 30 °C.

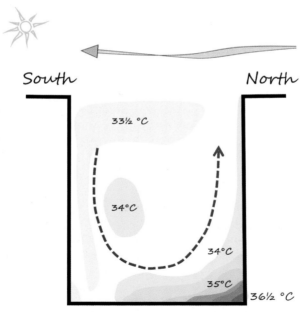

Fig. 2.18 The air temperature distribution within an east-west oriented street canyon for the period 1450–1500 h on 2 August 1983.
Modified from Nakamura, Y., Oke, T.R., 1988. Wind, temperature and stability conditions in an E-W oriented urban canyon. Atmos. Environ. 22, 2691–2700.

Fig. 2.19 Visible and thermal images of Atlanta (USA) from Landsat 7, 1000h on 28 September 2000. The urban area appears in shades of grey on the visible image (left) and in dark orange and red on the thermal image (right). From M. Jentoft-Nilsen, NASA. Source: https://earthobservatory.nasa.gov/images/7205/urban-heat-island-atlanta-georgia.

Studies confirm that the SUHI effect is greatest in clear, calm, and sunny weather and is larger during the day compared with the night, in contrast to the CUHI. During the daytime, T_s is extremely heterogeneous over short distances. Unlike T_a, which mixes the contributions of many different facets, T_s is linked intrinsically to individual characteristics of facets (that is, the cover, fabric, and geometry that regulate energy receipt, absorption, and withdrawal). The impacts of these characteristics are clearest in daytime when access to direct radiation accentuates the thermal response. At night, the facets cool and the differences diminish, revealing a simpler pattern closer to that of the CUHI.

As an example, Dousset et al. (2011) used TIR data from 50 satellite images to examine the temporal and spatial character of the SUHI at $1\,km^2$ resolution during an exceptional heat wave event in Paris (4–13 August 2003) that caused a public health crisis. At night, the SUHI was centered on the urban core with diminishing values toward the suburbs (similar to the CUHI), but during the day multiple peaks of SUHI were detected across the city. Detailed land-cover and land-use maps showed that these peaks were associated with large lightweight buildings with corrugated steel or fiber roofs (i.e., LCZ 8, Table 2.3). These buildings were used primarily for storage and were surrounded by extensive paved areas for transport and parking. After sunrise, the thin building envelopes warmed quickly, owing to their limited heat storage capacity and large roof area exposed to direct radiation. At night, these same surfaces cooled quickly owing to radiative loss (high SVF) and the small heat reservoir in the fabric, which was rapidly drained. Similarly, the outdoor paved spaces warmed and cooled quickly due to their low albedo, lack of vegetation (day), and large SVF (night). By comparison with these areas, the city center was warmer by day and by night during the heat wave period. The city center is densely built of stone and brick structures (LCZ 2, Table 2.3), which are occupied by residences and offices.

2.4.3   Comparing the CUHI and SUHI

One should be aware of the differences between the surface and canopy-level UHIs, both in terms of the processes responsible for each *and* how they are studied. This has important implications for the interpretation of UHI results and deciding on appropriate responses.

The urban effect on T_s would ideally account for the entire "skin" of the urban surface, that is, the walls, roofs, trees, ground, etc., with which the overlying air is in contact. This is the *complete* surface temperature. Practically, this is not observed at large scales and most SUHI studies use remote sensing techniques which involve directional instruments and record T_s only within a defined view that corresponds to a sample of urban facets. The satellite-observed SUHI has a "bird's eye" perspective of the surface and a spatial resolution that (1) integrates the contribution of a variety of surface types (roads, pavement, roofs, tree tops, etc.), and (2) emphasizes the role of visible horizontal surfaces at the expense of vertical surfaces such as walls. Critically, these instruments sample from facets inside and outside the UCL.

Satellite-based SUHI studies can provide synoptic observations of whole cities, overcoming many of the problems associated with studying the CUHI. However, the researcher has no control over the instrumentation and no opportunity to select the time of observation to correspond with weather conditions. The great advantage of these systems is the global coverage and their repeated observations over time. These attributes permit multicity comparisons of SUHI magnitude, and the examination of changes to SUHI properties over time as a result of rapid urbanization and/or shifts in weather/climate.

2.5   Concluding remarks

The emphasis on energy exchanges in this chapter provides a context for understanding the UHI. This perspective is important when preparing UHI studies (deciding on appropriate observation sites and routes), and analyzing, interpreting, and presenting the results. Moreover, it is needed if the study is used as a basis to support heat mitigation and/or adaptation polices, and subsequently to decide on the scales, places, and appropriate means of intervention. There are a couple of points worth noting:

- The urban effect can only be assessed by comparison with a benchmark site. Often this site is taken to represent the "natural" setting in the absence of urbanization, but other reference sites can be used. There are a great many types of urban and natural landscapes present in and around any city, and each type will have a distinct thermal effect on the surface and overlying atmosphere.
- The surface and canopy-level UHIs are linked but are the outcomes of different dominant processes. Moreover, the common methods for observing each are not correlated in space or time.

References

Dousset, B., Gourmelon, F., Laaidi, K., Zeghnoun, A., Giraudet, E., Bretin, P., Mauri, E., Vandentorren, S., 2011. Satellite monitoring of summer heat waves in the Paris metropolitan area. Int. J. Climatol. 31, 313–323.

Lowry, W.P., 1977. Empirical estimation of the urban effects on climate: a problem analysis. J. Appl. Meteorol. 16, 129–135.

Nunez, M., Oke, T.R., 1977. Energy balance of an urban canyon. J. Appl. Meteorol. 16, 11–19.

Oke, T.R., 1981. Canyon geometry and the nocturnal urban heat island: comparison of scale model and field observations. J. Climatol. 1, 237–254.

Oke, T.R., Mills, G., Christen, A., Voogt, J.A., 2017. Urban Climates. Cambridge University Press, Cambridge, UK.

Stewart, I.D., Oke, T.R., 2012. Local Climate Zones for urban temperature studies. Bull. Am. Meteorol. Soc. 93, 1879–1900.

Ward, H.C., Kotthaus, S., Järvi, L., Grimmond, C.S.B., 2016. Surface urban energy and water balance scheme (SUEWS): development and evaluation at two UK sites. Urban Clim. 18, 1–32.

Atkinson, W., 1912. The Orientation of Buildings: Or, Planning for Sunlight. J. Wiley & Sons, New York. Available at https://archive.org/details/orientationofbui00atki/page/n5/mode/2up.

UHI management

3

The urbanization of a landscape is a historical process that starts with the decision to select a site to establish a settlement. Once this decision is made, the background climate is set, which includes the typical pattern of synoptic-scale weather systems that regulate precipitation, wind, cloud, sunshine, etc. If the settlement is located in a complex topography, such as in a valley or next to a large water body, the climate will also be affected by regional-scale influences such as mountain-valley winds and land-sea breezes. In fact, even as the urban landscape is made, it covers natural undulations in the terrain and incorporates features like hills, valleys, rivers, and lakes, each of which has a local climate effect. Of course, the "natural" landscape around the settlement is also adapted, as forests are cleared, lands are drained, crops are grown, etc. At any given time, therefore, the urban climate is a product of natural and anthropogenic drivers acting at a hierarchy of spatial and temporal scales. This perspective is useful when considering the best way to manage the UHI, which represents just one aspect of the urban climate.

In Chapter 2, we placed the UHI phenomenon within an energy budget context that provides a process-based understanding of its causes. In this chapter, we focus on the policy responses to manage the UHI through actions designed to mitigate excess heat and/or adapt to its impacts. *Mitigation* addresses the causes of the UHI and modifies the urban landscape in response. *Adaptation* accepts the UHI as an inevitable outcome of urbanization and adjusts aspects of the city and human behavior to cope. Practically, these policies overlap at urban scales and are distinguished mainly by emphasis. Moreover, many of these policies are set in the context of global climate change (GCC) science, which predicts a warmer future (especially at night) and more frequent heatwaves. As a consequence, it is expected that, in cities, the combination of urban and global/regional warming represents a risk to public health. It follows, then, that managing the UHI is a component of broader initiatives to make cities more resilient and sustainable.

3.1   A UHI management model

A template for the scientific infrastructure needed to support UHI management could follow (and complement) those established for other environmental issues, such as air quality. It would have four linked components: observations of the phenomenon; understanding of its causes; limits/guidelines on impacts; and strategies to mitigate and/or adapt in response. Observations are needed to provide information on the types of UHI and where/when the effects are greatest. Understanding is based on the causative energy exchanges and may be incorporated into numerical models that can simulate

The Urban Heat Island. https://doi.org/10.1016/B978-0-12-815017-7.00003-5

the magnitude and timing of the UHI in response to changes in surface geometry, land cover, and land use. The observations are used to evaluate the model and improve its performance. Guidelines are required to assess the impact of the UHI on issues such as pollution, building energy use, and human health and well-being. The impacts should be measurable in terms of costs to health systems or energy production, for example. Finally, this scientific infrastructure is needed to support policy development, and to decide and evaluate the most cost-effective actions to address the UHI and its allied issues.

The basis for any UHI intervention is predicated on an assessment of risk, which has three components (Fig. 3.1). *Hazard* describes the statistics of the UHI including the frequency and impact of threshold events, such as heatwaves; *exposure* measures the extent to which people and/or infrastructure may be affected by the hazard; and *vulnerability* indicates the ability to cope with the consequences. To evaluate the need for interventions, we should consider the "when," "where," and "magnitude" of the hazard, in combination with "who" and "what" is impacted. For example, in Section 2.4.2, we presented the example of a satellite-observed SUHI for Paris during a heatwave event. During the daytime, high SUHI magnitudes were found in warehouse areas (LCZ 8), which consist of extensive paving and large buildings with thin envelopes. Although these areas experienced a strong SUHI effect, few people were exposed to the hazard. From a public health viewpoint, the risk here is relatively low.

Fig. 3.1 places the UHI risk within the context of climate and urbanization. The climate has a natural state with associated variability (such as heatwaves or droughts) of varying severity. This climate is being changed at global and regional scales as a result of emissions of greenhouse gases and changes to the Earth's surface. The emissions are produced mainly by fossil-fuel use, which is concentrated in cities. Many of these large-scale changes are enhanced at urban scales where intensive changes to land cover modify local energy exchanges. While it is mainly the wastes associated with urban metabolism that drive large-scale climate change, it is the replacement of natural cover that dominates changes at urban scales.

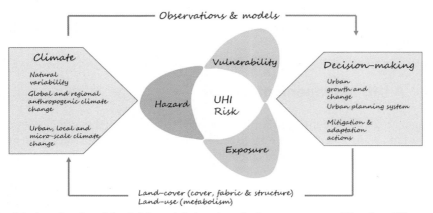

Fig. 3.1 An urban heat island risk model. Based on the Intergovernmental Panel on Climate Change (IPCC) risk model.

The geographic location of a city and its character, as expressed in its spatial extent and associated geometry, cover, and use, regulates its emission of greenhouse gases (large-scale anthropogenic climate change) and the surface-air energy exchanges (urban-scale climate changes). Modifying urban form and functions can address the various sources of climate change, but any actions must be coordinated to ensure that improvements at one scale do not have negative impacts at another scale. For example, the magnitude of the UHI can be reduced by decreasing built density and increasing the vegetative fraction, but this could increase the spatial extent of the city and its urban metabolism; in current urban systems, this will result in greater greenhouse gas emissions. By contrast, increasing built density by intensifying urban landscape change to create more compact and high-density cities will reduce the city's metabolic demands but enhance urban effects, such as the UHI.

3.2 Human climates

One of the main reasons for conducting UHI studies is to examine the consequences for human thermal health, both indoors and outdoors. The focus may be on the direct consequences of overheating, or on the building energy demand needed to offset excessive warming. Many UHI policies are designed to ameliorate its magnitude (hazard) or its effects on the population (vulnerability), and are often combined with strategies to address global climate change. For this reason, it is useful to discuss the basics of human climates, heat (dis)comfort, and its management. We use the energy budget framework introduced in Chapter 2 to discuss the processes responsible for regulating the body's temperature.

The energy budget for the human body can be expressed as (Fig. 3.2):

$$Q^* + Q_M = Q_H + Q_E + \Delta Q_S \tag{3.1}$$

The new term here is the metabolic heat flux (Q_M), which is generated internally at a rate controlled by one's level of activity. When at rest, $Q_M \approx 70\,\mathrm{W\,m^{-2}}$ but when engaged in intensive exercise, such as running, it can exceed $300\,\mathrm{W\,m^{-2}}$. Humans are homeotherms and seek to maintain a near-constant internal temperature. The body's thermoregulation system achieves this by minimizing the storage term (ΔQ_S) and by constantly adjusting each of the remaining terms to ensure balance (inputs = outputs). When this system is functioning, the body maintains a near-constant deep body temperature (T_b) of $37 \pm 2\,^{\circ}\mathrm{C}$.

Net radiation (Q^*) includes the receipt ($K\downarrow$) and reflection ($K\uparrow$) of shortwave radiation and the absorption ($L\downarrow$) and emission ($L\uparrow$) of longwave radiation. Direct (beam) radiation on the body depends on the location of the Sun in respect to the human body; when the Sun is directly overhead, one's head and shoulders receive the bulk of this receipt, but at higher zenith angles the burden shifts to the torso and limbs. The length of shadow generated is a good indicator of the intensity and distribution of direct radiation on the body: the longer the shadow, the less intense is the radiation receipt. Diffuse solar radiation is received from the sky hemisphere and from the surrounding

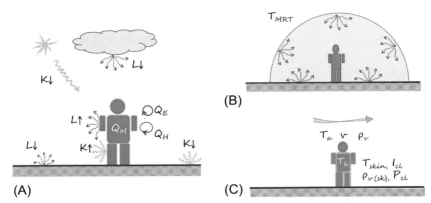

Fig. 3.2 The controls on the climates of humans. (A) The terms of the energy budget for humans, indicating the sources of shortwave (K↓) and longwave (L↓) receipt and loss (K↑ and L↑) on the body; the exchange of sensible (Q_H) and latent (Q_E) heat with the ambient atmosphere; and internal metabolic heat (Q_M) generation. (B) Radiation receipt on the body can be expressed as originating from a dome with a uniform mean radiant temperature (T_{MRT}). (C) The sensible and latent heat exchanges at the body surface are regulated by the ambient conditions (air temperature [T_a], air humidity [ρ_v], and wind velocity [v]); the properties of the skin surface (skin temperature [T_{sk}] and humidity [$\rho_{v(sk)}$]); and the insulation properties of clothing (I_{cl} and P_{cl}). The deep body temperature (T_b) remains near constant at 37 °C as long as energy inputs equal outputs.

environment by reflection. A proportion of K↓ is reflected: the albedo (α) of human skin is about 0.15, but for the clothed body this can be raised (light colored) or lowered (dark colored). Longwave radiation is diffuse in origin and is emitted from the sky vault and the surrounding environment based on their respective temperatures and emissivities. The body is a very efficient absorber of longwave radiation and reflects little. It follows that it is also an excellent emitter (ε ≈ 0.95).

Fig. 3.3 shows a visible and thermal image of street pedestrians taken with a thermal infrared (TIR) camera. It estimates surface temperature (T_s) based on the longwave radiation received from the features in the scene. The instrument applies a uniform value for ε, which is close to 1. The sky appears very cold due to its low emissivity and by comparison the other features (walls, street, and humans) are warm. Each of these has natural ε values close to 1. The effect of the shadows on the temperature of the walls and street is evident, and the relative warmth of the pedestrians is clear. This figure also shows the diverse sources of radiation receipt on the body, which have complex (and dynamic) geometric relations with the pedestrians. This makes the measurement of the radiation load (K↓ + L↓) on humans very difficult. To overcome this, we treat the radiation gain by the body as originating from a dome with a mean radiant temperature (T_{MRT}) and emissivity (Fig. 3.2B). The T_{MRT} can be acquired using a globe thermometer, which consists of a conventional temperature sensor within a spherical enclosure that has a known ε.

Sensible heat loss (Q_H) occurs through breathing and via the skin surface. Heat loss at the outer surface of the body is regulated by the skin temperature (T_{sk}) and ambient

Fig. 3.3 Visible (left) and thermal (right) images recorded with an infrared camera (assuming uniform emissivity). The images show a street partly in sunshine with pedestrians; the sunlit area and the humans are distinct and the sky appears as a very cold feature (owing to its low emissivity). The estimated surface temperature in the target area is 15.4 °C.

air temperature (T_a). Clothing alters the exchange at the skin by providing a layer of insulation (I_{cl}), which traps air against the skin surface and shifts the "active" surface for the body to the cloth-air interface. Mixing at this interface is linked to the ambient wind speed (v), which removes air next to the body. Higher I_{cl} values reduce heat loss from the body, while airflow enhances heat loss. In hot climates, high albedo clothing that is loose on the body and offers little insulation is best where winds are light.

$$Q_H = f\left(T_{sk}, I_{cl}, T_a, v\right) \tag{3.2}$$

Latent heat losses occur through breathing, regulatory sweating, and diffusion through the skin. At the skin surface, heat loss occurs by evaporation, the rate of which depends on airflow, humidity at the skin ($\rho_{v(sk)}$) and adjacent air ambient humidity (ρ_v), and the effects of clothing. The permeability of clothing (P_{cl}) can also be designed to limit heat loss through evaporation with the ambient air.

$$Q_E = f\left(\rho_{v(sk)}, P_{cl}, \rho_v, v\right) \tag{3.3}$$

Fig. 3.2C illustrates these exchanges and their relation to body, clothing, and atmospheric variables. Note that the exchange of energy, via conduction, is treated as negligible in these cases, as a standing person has limited direct contact with the ground.

The body's thermoregulation system consists of a series of biophysical responses to manage energy exchanges and maintain a near-constant temperature in the body core, which includes the center of the torso and a narrow corridor to the brain. These responses are a reaction to excessive heating (cooling) and are designed to limit (boost) energy production (Q_M) and enhance (limit) heat loss. One of the rapid response measures used by the body controls the flow of blood to the skin surface, allowing T_{sk} to change. When the body is comfortable, T_{sk} is about 33 °C (4 °C lower than the core), but this can be raised (vasodilation) and lowered (vasoconstriction) in response to warming and cooling, respectively. Managing the skin temperature alters Q_H, which

Fig. 3.4 Direct solar radiation is the main driver of outdoor human (dis)comfort, as illustrated by the preferred locations of people outdoors. Left: a park in a cool climate (Dublin, Ireland). Right: a plaza in a hot climate (Istanbul, Turkey).
From G. Mills.

depends on $(T_{sk} - T_a)$ and $L\uparrow$, which is regulated by T_{sk}. However, this response provides a limited capacity to deal with excessive and rapid heat loss and gain. If the body continues to cool, metabolic heat production (Q_M) can be increased to compensate for heat loss. Similarly, if the body is warming, Q_M and Q_H will be reduced. The most effective means of heat disposal is via regulatory sweating (Q_E), whereby water is extruded via sweat glands and evaporated. In addition to these internal mechanisms, humans make deliberate decisions to modify these energy exchanges through choices of clothing and/or moving to another environment where ambient conditions are either warmer (in sunshine) or cooler (in shade). While the range of ambient conditions to which humans can adjust is wide, there are limits to our ability to regulate temperature through these responses. Excessive heat loss that cannot be sustained through internal heat generation (Q_M) will eventually result in hypothermia, cooling of the body core, and death. Similarly, the body's cooling system (Q_E) will become overwhelmed if evaporation cannot offset heat gain, leading to hyperthermia, warming of the body core, and death.

UHI researchers interested in the design of spaces to create microclimates that people wish to occupy would do well to observe the use of space in response to weather conditions. Fig. 3.4 highlights the responses of people to outdoor conditions through the selection of preferred locations. It juxtaposes responses in cool and warm climates where individuals at rest chose to occupy areas in sunshine and shade, respectively.

3.2.1 Thermal stresses and strains

The psychological state in which the individual has no desire to be warmed or cooled is defined as *thermal comfort*; note that the definition includes personal judgement and is not simply an outcome of achieving energy balance. This judgement incorporates one's thermal expectations of the ambient environment indoors and outdoors at different times of the day and across seasons. Discomfort is simply a deviation from this neutral state, usually measured on a spectrum from hot to cold.

There is a considerable body of research on the links between discomfort and measures of environmental *stress* (due to meteorological conditions) and the body *strain* (e.g., skin temperature and sweat rate). This distinction is based on the conditions to which the body is exposed (stress) and its physical response (strain). From an applied climate perspective, it is the former that are commonly used to assess the ambient conditions that are suited for particular uses. Typically, these conditions are expressed using indicators that measure the net impact of ambient conditions in terms of an equivalent air temperature (T_{eq}) that would result in the same level of stress under standardised conditions. Meteorological indicators (such as heat indices) are designed for general use and employ observations of air temperature, relative humidity, and wind to provide public health guidance on the weather conditions likely to cause moderate to severe thermal stress. These indices are appropriate at urban scales but have limited diagnostic value as they do not help address the causes of the stress and cannot account for microscale variability. Place-specific assessments of stress rely on observations of the microscale climates to which people are exposed and include the effects of radiation (using T_{MRT}):

$$T_{eq} = f\left(T_{MRT}, T_a, \rho_v, v\right) \tag{3.4}$$

Much of the interest in UHI stems from a concern for enhancement of heat stress during warm weather, although it may also result in a reduction of cold stress at other times of the year. Periods of exceptional heat stress are often associated with meteorological events known as heatwaves. The WMO defines these as *a period of marked unusual hot weather over a region persisting at least three consecutive days during the warm period of the year based on local climatological conditions, with thermal conditions recorded above given thresholds.* Note that this describes heatwave in terms of a period when T_a values are "unusual" and above a given "threshold"; in other words, the precise criteria depend on the climate at a place. This definition recognizes that societies are acculturated to their normal climate and that this is reflected in the design of buildings, outdoor spaces, and behaviors. Nevertheless, the synoptic weather conditions associated with heatwaves are generally the same and are associated with high pressure systems, the advection of warm (and sometimes humid) air in light winds, and strong solar radiation. Heatwaves represent a great risk to public health and periods of exceptional warming are linked to excess deaths associated with hyperthermia. The impact of these events is concentrated in vulnerable communities, which often correspond with poor and elderly populations.

3.2.2 Managing the human climate

The majority of this thermal work has historically been completed primarily to establish the indoor climate suited to types of work. In these conditions, it is possible to limit wind speed and ensure $T_{MRT} \approx T_a$, so that "comfort" can be ensured by managing just temperature and humidity, which are maintained at a near-constant level throughout the working day. Increasingly, research on human thermal (dis)comfort has addressed outdoor spaces. Although these spaces are generally much more difficult to manage than indoor spaces, applying the same principles can address causes of discomfort at critical times/periods. Indoor and outdoor climates are thermally separated to varying degrees but decisions on indoor and outdoor spaces affect each other.

The climate of a building is managed by regulating the flows of energy across the envelope (external loads) and the generation of energy internally by lighting, cooking, computers, people, etc. (internal loads) (Fig. 3.5). The role of a heating, ventilation, and air conditioning (HVAC) system is to balance these loads to ensure a desirable indoor climate. The energy demand of a building is closely tied to its purpose (for example, business or residential) and its occupation patterns (daytime, weekday, etc.). Most buildings are designed for the comfort of the occupants, but the energy needed to accomplish this depends on the operational definition of comfort and the extent to which it deviates from the ambient conditions outdoors. In highly managed indoor environments, a *thermal comfort model* may be adopted that applies a strict set of ambient conditions based on universal criteria, and the role of the HVAC system is to maintain these. For a typical office building, this climate is measured using air temperature and relative humidity; common indoor values would be set at $T_a \approx 22\,°C$ and RH $\approx 50\%$, with minimal air motion. In effect, this type of building is designed for the occupants to wear similar clothing throughout the year, and to provide a climatically neutral environment that is thermally divorced from the outdoors. Alternatively, applying an *adaptive comfort model* permits variation in the indoor climate within an acceptable range that is linked to seasonal variations in the outdoor temperature. In this type of building, the occupant takes responsibility for much of the indoor climate control through opening/closing windows and wearing clothing appropriate to the time of year. Finally, a free-running building creates an internal climate without

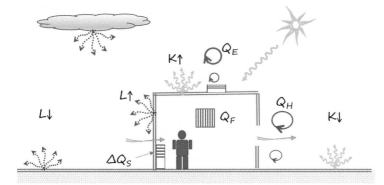

Fig. 3.5 The energy budget terms for the simple occupied building are given, including shortwave (K $\downarrow$) and longwave (L $\downarrow$) receipt and loss (K $\uparrow$ and L $\uparrow$); the exchange of sensible (Q_H) and latent (Q_E) heat at the envelope; and internal heat generated by lighting and cooking, and by heating, ventilation, and cooling systems, etc. (Q_F).

the need for an HVAC system or the need for any internal energy source to offset excessive heat gain and/or loss.

Many of the assumptions and approaches that underpin research into climate indoors are not applicable to outdoor settings where the ability to manage environmental stresses is much more limited. Moreover, unlike indoor settings, an individual is exposed to constantly changing (transient) ambient conditions and is perpetually adjusting. Of course, the thermal expectations of those outdoors are far less restrictive and decisions on clothing and preferred settings depend on the type of activity, such as sitting or walking or running. In hot weather conditions, the major determinants of daytime heating stress are air temperature, relative humidity, wind speed, direct radiation from the Sun, and the total radiation gain from the environment. Frankly, it is not possible (or even desirable) to ensure comfort across the urban landscape but it is possible to create spaces that have microclimates suited to outdoor uses at different times of the year (Erell et al., 2012). This requires managing the suite of energy gains experienced by humans that, in combination, cause heat stress. This may mean providing shade, reducing T_{MRT}, enhancing evaporative cooling, and encouraging ventilation. Of these variables, T_{MRT} is perhaps easiest to manage at a microscale (< 100 m) through changes to surface cover and tree planting to cast shadows. Strategies to increase near-surface wind speed (v) involve changing aspects of urban geometry to reduce obstacles to airflow and may require neighborhood-scale (> 1 km) alterations. Altering air temperature and humidity directly at scales > 100 m is difficult given the atmosphere's mixing ability, but it is possible at smaller scales (< 10 m). Note that the body's thermoregulatory system responds to the integrated effects of environmental stresses, rather than any one variable.

3.3 Heat mitigation

The causes of the UHI can be attributed to changes in:

- structure—the dimensions and layout of buildings that create a surface geometry
- cover—the replacement of natural cover and vegetation with paving;
- fabric—the use of manufactured materials with distinct thermal and radiative properties;
- metabolism—the intense occupation of space and the waste energy emitted.

The net effect of these changes is to alter the urban energy budget at a series of scales and warm the surface, the overlying air, and the underlying substrate. To state the obvious, mitigating urban heat requires altering current urbanization practices so as to reduce heat gain and enhance heat loss. Altering the surface energy balance (and temperature) of a given urban feature (a road, a roof, etc.) is relatively easy, but modifying T_a in and around buildings is more difficult because of mixing. Stating the objectives of the intervention at the outset is a good way of ensuring that the intent is matched to scale (building, street, and neighborhood) and desirable outcome (e.g., building energy use, air quality, and heat stress). Before discussing UHI policies, we will discuss UHI management in the context of the energy budget framework and focus on daytime warming and nighttime cooling.

3.3.1   Regulating daytime warming

The primary driver of surface warming during the daytime is the absorption of short-wave radiation (K ↓ - K ↑), which heats up the overlying air (Q_H) and underlying substrate (Q_G); by comparison, net longwave radiation (L ↓ - L ↑) is relatively constant through the day. For shortwave radiation at a surface, the options are to increase reflectivity (albedo) and/or provide shade. For the near-surface air, it is to reduce the turbulent sensible heat flux (Q_H) and enhance the latent heat flux (Q_E). For the substrate, it is to manage the conductive heat flux (Q_G) and the energy held in storage (ΔQ_S).

3.3.1.1   Radiation

Albedo can be changed by replacing the fabric itself (light-colored concrete versus darker asphalt) or simply adding a reflective coating such as white paint. This type of coating reflects in the visible spectrum and can cause visual glare. More sophisticated materials reflect in the near-infrared or even change their reflective properties based on surface temperature (thermochromatic). The latter are ideal when low-albedo materials are desirable in colder weather to increase solar absorption and surface warming, and high-albedo materials in hotter weather to do the opposite. The application issue is to consider which surfaces (facets) can be altered and the implications for neighboring facets and outdoor users that will receive the reflected radiation. Facets that are designed to have high albedos are termed *cool surfaces.*

In ideal weather for strong UHI formation, the sky is clear and most (> 80%) of K ↓ consists of direct (beam) radiation, which has a singular origin. Given that the position of the Sun by time of day/year is immutable, strategic shading is one of the most effective means of limiting daytime heating at a surface of interest. Keep in mind, though, that the total amount of direct radiation received at the Earth's surface remains the same and that shading simply transfers this energy from one surface to another. Patterns of shading are dynamic and are a function of solar position (zenith and azimuth angle) and the geometry of the surface features; in cities, street patterns exert a strong control on the timing and extent of shading within the urban canopy layer.

Managing diffuse (short- and longwave) radiation receipt at a surface of interest is more difficult. Within the UCL, the two general sources of diffuse radiation are the sky and the surrounding urban landscape; the contributions of each depends on the magnitude of the radiant source and the sky view factor (SVF). Diffuse solar radiation from the sky is an important source of natural daylight that can offset some of the lighting needs of buildings, but it is not a significant driver of daytime heating. Similarly, longwave radiation from the sky source does not vary much through the daytime, so regulating it will not influence the urban warming effect. In other words, it is the radiation exchanges within the canopy (Fig. 2.13) that are most amenable to managing daytime surface warming. However, remember that in the calm and clear meteorological conditions associated with strong UHI events, the dominant driver is direct (beam) solar radiation.

Increasing the albedo of wall and street surfaces will simultaneously enhance shortwave receipt and reduce longwave receipt at other surfaces within the UCL. The net impact will be to "spread" radiation receipt more evenly within the canopy and reduce the average surface temperature. The magnitude of this effect will depend on the geometry of the UCL and the direct radiation receipt. For the example of a simple street, the height-to-width ratio (H/W) is an indicator of the SVF; a H/W value of 1 has a SVF of 0.5 at the center of the street surface. A wider street (H/W < 1) allows more solar radiation into the canopy but there will be less surface-to-surface exchange; a narrower street generates more shade but increases these exchanges. Clearly, the ideal geometry from the perspective of managing surface temperatures within the canopy depends on the solar energy gain, which controls the energy available for recycling within the UCL. In hot climates where the Sun has a small zenith angle for much of the year, designing narrow streets to limit daytime energy gain is a sensible response, as it will limit surface heating within the UCL even if there is more recycling. In cooler climates, where hot weather events are seasonal and street geometry must strike a balance between solar access and shade, wider streets are better suited. In these circumstances, trees provide the best solution to this conundrum, especially deciduous trees that shed leaves during the winter months when solar access is desirable.

3.3.1.2   Turbulent heat exchanges

Available energy at the surface ($Q^* - \Delta Q_S$) is transferred to the overlying air by sensible (Q_H) and latent heat (Q_E) fluxes (Eq. 2.7). The magnitudes of these exchanges are functions of the energy available, the surface-air gradients of temperature and humidity, and atmospheric turbulence. The options for controlling surface and air warming are to enhance Q_E at the expense of Q_H and/or to enhance turbulence so that heat is mixed into a deeper layer of atmosphere, thereby diminishing the temperature impact.

The Bowen ratio ($\beta = Q_H/Q_E$) indicates the preferred route of exchanges such that for $\beta < 1$, energy is expended on evaporation rather than direct heating. The daytime canopy-layer and boundary-layer UHIs are partly the result of enhanced Q_H ($\beta > 1$) because the urban surface is generally dry and lacking in vegetative cover (Fig. 2.6). Increasing water availability either directly (increasing water cover) or indirectly (increasing vegetative cover) will enhance the surface-air humidity gradient, divert more of the available energy to evaporation, and reduce β (Fig. 2.16). These landscaping tools are referred to as blue (water) and green (vegetation) infrastructure, respectively, and the design considerations are the extent and location of the urban surfaces to wet and/or vegetate. A more complex response is to enhance the mixing of air; while this will increase the magnitude of the turbulent exchanges, it will spread heating into a large volume so that the temperature increase is muted. This could be accomplished by the strategic placement of tall obstacles (including buildings) to generate near-surface turbulence. At the microscale, mechanical fans provide the same function and are often used in outdoor sitting areas to cool people.

3.3.1.3   Heat storage (ΔQ_S)

Typical paving materials are efficient stores of sensible heat owing to their relatively large thermal admittance values (due to high conductivity and high heat capacity). This is largely a product of the density of materials needed to support the weight of vehicles and protect substrate infrastructure. Permeable paving systems are designed mostly to manage urban runoff, but they will also have thermal effects as they permit the underlying soil to receive and lose water, allowing their thermal properties to vary with the weather, much like natural soils. Moreover, as these systems provide air-spaces for the movement of water, when dry, the admittance of the fabric is lowered and heat storage is reduced. Building walls made of stone and brick have much the same effect as paving; modern buildings that have thinner envelopes with insulation to limit conduction to the building interior can greatly reduce ΔQ_S. Keep in mind the principle of energy conservation: limiting heat storage only, without managing solar absorption, simply means that more of the energy available at the surface is used to raise surface temperature and enhance the turbulent fluxes. It will, however, reduce the energy store that is available at night for withdrawal.

3.3.1.4   Anthropogenic heat emission (Q_F)

The addition of anthropogenic heat to the urban atmosphere occurs at a variety of scales and is typically divided into building, transport, and manufacturing sources. The magnitude and geography of these emissions will depend on the economy of the city and its land-use distribution. The building contribution can be attributed to waste heat from two sources: (1) the day-to-day functions that require hot water, computers, lighting, etc.; and (2) the heating and cooling needed to provide a comfortable indoor climate. While the former varies with time of day and day of week, the latter is driven by outdoor climate. Q_F from this source can be managed by reducing energy demand (e.g., choosing a different comfort model) or energy use (e.g., improved building insulation). Waste energy from buildings is conducted through the fabric and exits by infiltration and via exhausts. From the vantage point of the UCL climate, exhaust outlets on walls should be discouraged in favor of roof locations. The same principles of efficiency and demand management apply to other anthropogenic sources such as transport; for example, improving vehicle efficiency (e.g., by electrifying much of the fleet) and designing mixed-use neighborhoods that reduce the need for private vehicles will have the same effect.

3.3.2   Regulating nighttime cooling

After sunset, the Earth's surface cools down by longwave radiative heat loss ($L\uparrow > L\downarrow$), which is supported by the withdrawal of heat from storage (ΔQ_S). Under the clear and calm conditions associated with strong UHIs, there is little turbulence so Q_H and Q_E are very small (Eq. 2.12). The options for managing the UHI at night, then, are to limit daytime heat storage, maximize longwave radiation loss, and enhance turbulent exchanges. The thermal memory in the near-surface climate system is ΔQ_S and, if there is less energy stored during the daytime, there will be a smaller reservoir to draw down at night. Hence, methods to reduce surface-energy gain during daytime (such as increased albedo or shade) will affect the "starting position" for nighttime cooling.

Fig. 3.6 People sleeping on building rooftops in Jaisalmer, Rajasthan (India). From Bazzano Photography/Alamy Stock Photo.

Longwave loss at a surface under clear-sky conditions is regulated by the sky view factor. Where a surface is flat and has an unobstructed view of the sky, SVF equals unity and radiative heat loss is maximized. In these circumstances, the surface cools exponentially after sunset, rapidly cooling at first and then more slowly. Where a surface is corrugated, different surface facets will have obstructed views of the sky (SVF < 1) and cool more slowly. As a result, roof surfaces will be cooler at night than either the walls or street of the UCL. This is apparent in Fig. 3.6, which shows people making use of the rooftop as a more comfortable sleeping area during hot nights. Changing street geometry simply to enhance longwave radiation loss is not really a viable option. Similarly, enhancing street level turbulence to generate mixing using mechanical fans (much like the systems used to prevent frost formation in orchards during cold, clear, and calm weather conditions) would also diminish the canopy-level UHI , but this is only a viable option in specific settings. Simpler design solutions could offer some limited respite: for example, in places where dwellings have flat roofs and are used as cool nighttime spaces, cooler air could be displaced into the street below using drainage channels.

3.4 Heat adaptation

Adaptation strategies do not directly address the causes of the UHI but deal with its consequences. This means changing urban infrastructure and human behavior to cope with higher temperatures. In its simplest form, adaptation would mean changing patterns of behavior to match the warmer weather, including altering clothing; modifying exercise, work, and sleep patterns to reduce metabolic heat generation (Eq. 3.1); and seeking cooler spaces to occupy, such as shaded areas or air conditioned buildings.

Cooling indoor spaces using air conditioning systems is a conventional response to hot weather in many cities. However, unlike other forms of adaptation, the net effect is to increase the anthropogenic heat flux (Q_F) to the outdoors (Fig. 3.4A), adding to urban warming. During exceptional heatwave events, many cities use accessible air-conditioned buildings as a means of addressing public health concerns for vulnerable residents. Other types of adaptation can also be seen as mitigation actions, as they seek to counteract the urban temperature effects, even if temporarily. Outdoor shading and small-scale water features (pavement wetting, fountains and sprays) can be implemented strategically in hot weather and specific circumstances to manage extreme heat (Fig. 3.7). More generally, water and vegetation can be used as fixed design features to mitigate the urban temperature effect in specific areas; if they are seen as part of an adaption strategy, they need to be accessible to those seeking thermal relief.

The UHI effect extends through the lower atmosphere to a depth of 1–2 km during the daytime and will result in a thermally driven circulation. This circulation is initiated by the formation of a weak low-pressure system above the urban canopy layer that draws air from the surrounding landscape. This air rises slowly over the city center, expands outwards at height and descends some distance outside the city, closing the circulation. Some researchers have suggested that the near-surface circulation (known as a country breeze) into the city could be encouraged by establishing wind corridors that channel cooler "rural" air into the urban area. While this idea may be fanciful, cities can make more effective use of natural topographic variation and land-use management to both mitigate and adapt the UHI at neighborhood and city scales (Fig. 3.8).

Fig. 3.7 Water and shade are effective "tools" to make outdoor spaces more comfortable for use. Left: a pedestrian zone in Nimes (France) enhanced by trees and small pools. Right: a commercial street in Marrakech (Morocco) covered by canvas.
From G. Mills and M. O'Connell.

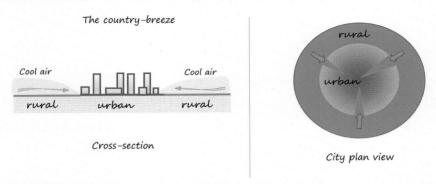

Fig. 3.8 Urban-scale land-use management can utilize local circulations to channel cooler air into the city along corridors that offer little resistance to near-surface air movement. Here a weak circulation oriented toward the city center and caused by the UHI is directed along linear parks.

3.4.1 Local wind systems

The UHI in many cities can be partly addressed by incorporating local wind systems into urban planning and design. These circulations are generated by topographic variations associated with orography (e.g., hills and valleys) and coasts. Globally, a great many cities are situated on coastal plains, in river valleys, and adjacent to mountain slopes, and therefore experience these wind systems. During conditions that are ideal for the formation of strong UHIs (calm and clear regional weather with strong direct radiation), these topographies generate thermally driven circulations at a variety of scales that can advect cooler air into cities and offset the urban thermal effect.

During daytime along coasts, the differential heating of land and water creates uplift over land and a flow of cooler and more humid air from the sea. This is known as a sea breeze. The intensity of the breeze depends on the magnitude of the land-sea temperature difference, and its extent depends on the size of the water body. By late afternoon on a hot day, the sea breeze can extend several kilometers inland as a shallow layer of sea air over a coastal plain; however, its progress can be impeded by obstacles along its path. In daytime UHI conditions, the arrival of the sea breeze is associated with an increase in windspeed and a drop in air temperature. At night the circulation reverses and a land breeze develops, moving air near the surface toward the ocean. For coastal cities, the sea breeze provides a natural air-conditioning system, which can eliminate the canopy-level UHI in affected parts, beginning at the coast and extending inland. To make best use of this breeze, the urban layout should be arranged so as to encourage this flow, with wide streets that are perpendicular to the coastline and free of large obstacles (Ng, 2009; Fig. 3.9).

Similar flow patterns occur on slopes under ideal heat island weather conditions. During the daytime, heating of sloping surfaces causes warming that draws air upslope (anabatic), whereas at night, surface cooling causes cooler, denser air to move downslope (katabatic). In complex terrain, these flows can become arranged into valley (upslope) and mountain (downslope) circulations by day and night,

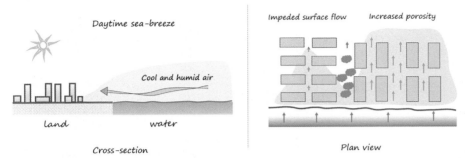

Fig. 3.9 The daytime sea breeze forms under regionally calm and clear conditions when land-sea surface temperature differences are large. By mid-afternoon, the breeze is established as a flow across the coastline bringing cooler air inland (cross-section). Building layout along the coast can impede (or encourage) the passage of this cool air into the city (plan view). During warm periods, the sea breeze can act as a natural air conditioner for the city.

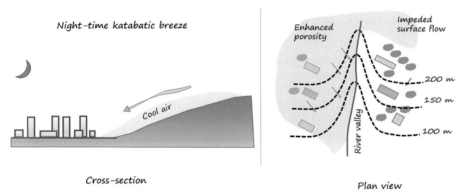

Fig. 3.10 Downslope (katabatic) winds form at night under clear skies and regionally calm weather conditions. The winds bring a shallow layer of cool (and clean) air into the city (cross-section), which can be encouraged through urban land-management that limits building on slopes (plan view).

respectively. Orographic circulations can be quite shallow (< 100 m in depth), but the katabatic flows in particular have been identified as important natural features that can be incorporated into city planning. In these circumstances, preserving elevated and exposed hill/mountain tops where cooler air is produced, as well as river valleys as passages for katabatic flow, can provide a network for bringing cooler air into the city (Fig. 3.10).

The Urban Climate Map process is designed to incorporate these local climate features into urban planning and design. As each city is uniquely situated, the process is based on mapping the urban area into climate types (climatopes) to represent the relation between topography, atmospheric properties, and their value for settlement design.

3.5 Heat management tools

The best mix of planning/design tools to manage urban heat depends on the character of the city and its background climate. Of course, the selection of tools and the scales of implementation should be based on the temporal-spatial properties of the UHI and its response to weather and climate, current and projected (Rosenzweig et al., 2018). Mitigation and adaptation strategies can be organized by scale into buildings, streets, neighborhoods, and cities (Table 3.1).

3.5.1 Buildings

Buildings are a focus of heat mitigation/adaptation for obvious reasons: they represent a large proportion of the urban surface, especially in compact neighborhoods (Table 2.4); they are important sources of anthropogenic heat (Q_F), much of it used to offset the outdoor climate; and their occupants/owners have a vested interest in energy management. There has been a particular emphasis on the roof facet and reducing its surface temperature during hot weather. In many cities, roofs may be flat, covered by dark materials, and poorly insulated. As a consequence, heat absorbed at this facet is readily conducted into the building space below, contributing to high indoor temperatures. In the analysis of heatwave deaths in urban areas, the populations that occupy the space next to these roof surfaces are identified as vulnerable, especially if they cannot afford cooling systems. Roof interventions focus on materials/coatings with high reflectivity (cool roofs) and green cover. Cool roofs will reduce the absorption of solar radiation, which is the main driver of daytime temperature extremes.

Green roof systems are used to channel available energy to evaporation rather than direct heating, but they are also used to harvest and store rainwater. *Extensive* systems consist of thin soil layers that can support ground cover involving hardy plants. These systems can also be grown on platforms that are displaced from the underlying building surface so they provide shade. *Intensive* green roof systems must be placed on near-horizontal roof surfaces, and have deep soil layers that can be used for planting trees. The soil acts as a layer of insulation and a store of water. Suitable roofs are very strong and the space may be used as a roof garden for building users.

The walls of buildings have received less attention for heat mitigation. Green walls describe vegetation that is fixed to the wall itself or to a supporting frame, which is attached to the wall (Fig. 3.11). These can provide some of the same services as the extensive green roofs but have the additional benefit of direct contact with the air below roof level. Green walls will generally have a lower temperature than uncovered walls and can be used to reduce the mean radiant temperature (T_{MRT}) experienced by people outdoors. More often, however, green walls are used as part of a strategy to improve air quality at street level.

Adaptation at the building scale includes responses to counteract climate-induced energy use (air conditioning) and/or changes to occupant behavior. The latter could mean relocating within the building to more comfortable spaces, or employing an adaptive comfort model that relies on the occupant to open/close windows, employ shading devices, etc. Many "smart" buildings rely on mechanical systems to make

Table 3.1 Actions on aspects of urban form and function (rows) at each urban scale.

Attribute|scale	Buildings	Streets	Neighborhoods	Cities
Structure Manage net radiation fluxes and near-surface wind velocity.	Design to minimize external energy gains and enhance heat loss through natural ventilation.	Design to manage daytime heat gain and nighttime heat loss. Vary geometry to ventilate canopy layer.	Urban layout to limit solar gain and increase porosity to airflow.	Higher building and occupation density (compact cities). Design with topography to channel natural ventilation.
Cover Enhance evaporation and water storage. Limit direct heating.	Green roofs and walls to provide a layer of insulation and evaporative cooling.	Tree planting in paved areas to provide shade and cooling for buildings and pedestrians.	Accessible green parks and gardens (lakes and ponds) to provide cool spaces for adaptive purposes.	Increase green/blue infrastructure at all scales. Make use of natural systems like river channels as linear parks.
Fabric Control heat absorption and storage.	Select materials that have higher albedos, especially on roofs. Glazing with reduced transmissivity.	Cool paving with higher reflectivity.	Permeable paving cover for pathways and parking areas.	Lower fraction of impervious surface cover.
Metabolism Reduce resource use and emission of waste heat.	Reduce energy use by managing demand (occupancy patterns and comfort levels) and improving efficiency of heating & cooling systems.	Reduce motorized traffic and shift to electric vehicles.	Mixed-use development to reduce traffic. District heating & cooling systems to reduce energy use.	More densely occupied cities to reduce per capita energy use. Renewable energy generation.

The attributes of form are structure (geometry), cover (vegetation), and fabric (material); together these alter the natural energy exchanges (Q^*, Q_H, Q_E, and ΔQ_S). Metabolism describes resources needed to support urban functions that result in waste heat (Q_F). Actions at building and street scales, if repeated, will affect neighborhood and city outcomes. Actions at street, neighborhood, and city scales have wider public benefits. Actions at building scales largely affect the occupants and can be supported by policies focused on energy management.

Fig. 3.11 Different approaches to managing building climates and indoor-outdoor exchanges. At left, the office building has rooms that are naturally ventilated and others that are air conditioned. At right, the hotel building uses vegetated balconies to provide shade and cooling for its guests.
From G. Mills.

these adjustments based on the readings of environmental sensors. If these systems employ renewable energy resources, then they can be considered a mitigation action for large-scale climates.

3.5.2 Streets

At the street scale, geometry exerts considerable control on all components of the surface energy balance. In a simple street canyon, lowering the aspect ratio (H/W) by lowering the heights of buildings and/or widening streets reduces the area of daytime shadowing and the contribution of multiple reflections; it also increases the SVF and enhances longwave radiation loss at night. In other words, decreasing H/W increases daytime surface heating and nighttime cooling. Increasing H/W has the opposite effect on radiation exchanges and also increases the sheltering effect, reducing mixing in the UCL. The "best" aspect ratio depends on the climate and the relative importance of access to sunlight and of shade. In hot and arid climates characterized by high daytime and low nighttime temperatures, very narrow streets provide cool, shaded spaces during daytime and warm, sheltered spaces at night, thereby counteracting the natural climatic stresses. Street orientation affects the distribution of radiation energy on walls and the ground over the course of a day. Orthogonal street patterns result in one side of the street getting sunshine during the day. For north-south streets, the side in sunlight switches from morning to afternoon but for east-west streets, one side is nearly always in shade. Streets that are oriented at other angles exhibit a more even distribution of solar radiation during the day/year (Fig. 3.12). Although orientation has little impact

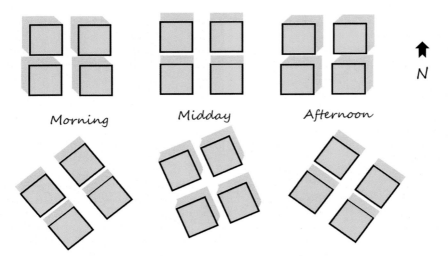

Fig. 3.12 Plan view of two clusters of four buildings oriented on orthogonal and off-orthogonal grids. The relative position of the Sun during the day produces distinct street-shadow patterns.

on the canopy-level UHI, it does impact the surface UHI (and T_{MRT}) at a microscale, and also has impacts for the use of outdoor spaces in hot weather.

Decisions on geometry are best made when the urban landscape is being created—changing street orientation and altering building heights and spacing after the fact is not usually an option. At street level, there are various options for temperature control; however, the choice (and impact) depends on the use of the street and the space available for actions:

- Cool paving raises the albedo and reduces solar radiation absorption, but unlike roof surfaces, the issues of glare and displaced energy to pedestrians outdoors must be considered. Moreover, on trafficked streets the cool surface will become dulled over time.
- Permeable materials are designed to allow water to penetrate the surface into the soil substrate by providing gaps in the fabric. Commonly, it consists of both a frame constructed of manufactured materials to provide strength with gravel/soil in the intervening spaces. These surfaces may be partially vegetated and allow some evaporation. Sections of the street that are designed as gravel-covered "pits" are often described as rain gardens. These systems are designed for hydrological control but have the advantage of lowering surface temperature.
- Water can be an effective means of cooling at a microscale and can be used to wet surfaces or as part of a feature. Where water is readily available, wetting streets at critical times (such as prior to noon) will cool the underlying surface as available energy is directed to evaporation. This is especially the case when the relative humidity is low.
- Greening using extensive systems is sometimes an option, particularly where other uses are not intensive. Trees are the most versatile of tools as they provide shade to walls and paved areas, and cool air via evapotranspiration. Streets are harsh environments for tree growth (poor air quality, long periods of shade, confined root space, etc.) and appropriate "urban" trees need to be selected. Fig. 3.13 shows a green corridor that is also part of a light-rail transport system.

Fig. 3.13 Green "infrastructure" alongside conventional transportation infrastructure. Left: vegetation separates vehicular traffic from pedestrian spaces (Lyons, France). Right: trees line a light-rail system embedded in a grass surface (Milan, Italy).
From G. Mills.

Much of the anthropogenic heat flux (Q_F) in cities is concentrated in streets, where the heat added by buildings on either side is supplemented by vehicular emissions. One liter of petrol ($0.001\,m^3$) is the equivalent of 9.1 kWh of energy; thus, a car travelling 1 km along a city street at a speed of $20\,km\,h^{-1}$ would use the equivalent of 2300–3600 W (heavy vehicles moving at the same speed would use more than ten times), most of which is deposited into the atmosphere. Improving engine efficiency and reducing traffic (or switching to electric vehicles) will reduce Q_F and help to mitigate canopy-level heat. The main environmental reason for transport management is to improve air quality, but reducing traffic can permit changes to much of the paved landscape, including the street surface, parking spaces, and car parks.

3.5.3 Neighborhoods

Mitigation actions at buildings and streets will affect the neighborhood-scale UHI if repeated across the landscape. Table 2.4 contains the typical land-cover parameter values for Local Climate Zones (LCZ), each of which is associated with a distinct thermal response. The strongest daytime SUHI and nighttime CUHI will generally be located in compact LCZ types, which are distinguished by high building densities and impervious surface fractions. The compact types are distinguished from each other by the heights of buildings, which affect the SVF and sheltering at street level. We can use these data to consider the potential impact of mitigation at this scale:

- Between 40% and 70% of the plan area of compact LCZ types are roof facets. Lowering roof surface temperature will directly affect the overlying air but the impact on the canopy-level air temperature depends on reduced building energy demand and mixing of the above and below-roof air.

- On the ground, apart from the compact buildings, much of what remains is paved and impervious (between 20% and 60%, depending on the built fraction), and much of this is road space. Altering this surface has a direct effect on the UCL climate.
- The role of vertical wall facets on the UCL increases with building height; in compact highrise neighborhoods, these facets can make up more than 50% of the urban surface. Altering wall properties directly effects the UCL climate.

When deciding on interventions at this scale, one should consider the intended outcome. For example, if it is the climate experienced by those outdoors that is of interest, more than likely this corresponds to a layer of less than 5 m adjacent to the ground. If it is to address overheating in buildings, then it is the building envelope that is relevant and, although the roof may occupy a large fraction of the neighborhood plan area, if the buildings are very tall, rooftops will represent only a small part of the envelope.

Increasing neighborhood tree-cover as a greening strategy can also be an effective means of managing surface and air temperature. Trees are especially versatile as a design tool because they can be planted in paved areas, and the variation in species provides considerable selection-freedom in terms of height, volume, canopy shape, and so on. Their climatic attributes are associated with the elevation, size, and leaf density of the canopy, where the direct interactions with the atmosphere take place. Trees shade and cool the ground in their vicinity and evaporate water through the leaves, cooling the air that passes through the canopy. Large, healthy, and mature trees (> 30 years old) can be over 20 m tall, but most street trees will be less than 10 m tall, so their influence is confined to the first few floors of buildings (depending on placement) and the paving under the canopy. During the daytime, trees can be employed as effective shading "devices" to reduce the T_{MRT} in occupied places. At night, however, the leafy canopy will obstruct part of the sky vault, reduce the SVF, and hinder surface cooling in the vicinity. Therefore, from the perspective of people outdoors, trees that are designed to reduce daytime heating will also reduce nighttime cooling; however, the top of the tree canopy (which may still be within the UCL) will be cool by day and night.

Large areas of water and green parks at a neighborhood scale can be seen primarily as adaptation tools, as they can be used to create more comfortable spaces to which people can relocate during hot weather. During the daytime, lakes and green parks appear as cool "islands" within the UHI. For water bodies, the high heat capacity and ability to transmit, mix, and evaporate freely means that they are efficient stores of energy and do not exhibit strong temperature responses. So, while they will appear cool during the daytime, they may appear warm at night. By comparison, green parks consisting of low vegetation will be cooler during the day (evaporation) and at night (high SVF). There is some evidence that these cooling effects will extend beyond the park boundaries to a limited distance, offsetting some of the warming in neighboring areas. This influence may be aided if the park is large and is sloped so that cooler air is directed into parts of the city. However, for most small urban parks their main value is to provide comfortable microclimates. For this purpose, the ideal park will be landscaped to offer areas of shade and shelter alongside open spaces. If an area is redesigned to remove paving and buildings to make a green park (change of use and cover), then that could be seen as a component of heat mitigation; however, once in place, the park is best seen as an adaption action, providing a desirable microclimate for those nearby.

3.5.4 Cities

The classic canopy-level UHI map that displays the temperature increasing toward the urban center is a result of historic urbanization patterns, in which case compact neighborhoods (LCZs 1–3) are located in the center and open neighborhoods (LCZs 4–6) on the urban outskirts. Consequently, moving from the edge to the city center brings increased paving, closer and taller buildings, and higher air temperature, especially at night. The SUHI has a more complex pattern in the daytime and the highest values are likely to be recorded over dark-colored paved surfaces or in warehouse areas (LCZ 8). These neighborhood types can be located closer to the urban edge than the center. To a considerable extent, then, the magnitude of the UHI can be mitigated by land-use and land-cover management. Simply put, the UHI would be reduced by increasing the area of open settlement patterns (LCZs 4–6) at the expense of compact patterns (LCZs 1–3). Clearly, this is not an option for most cities.

It may be possible to adapt the city to its climatic context by making best use of topographic variations that produce local to city-scale circulations during the same weather conditions that generate strong UHI events (Fig. 3.10). To make best use of the flows, urban design needs to ensure that there are few obstacles to movement along preferred channels to advect cooler air into the city (Fig. 3.8).

3.6 Costs and benefits

Decisions on UHI management are typically framed in terms of a cost-benefit analysis (CBA) using measures of value, often expressed in monetary terms. The costs describe the expense and effort associated with actions to alter the urban landscape to mitigate and/or adapt to the heat island effect. The benefits describe the impacts of regulating excessive urban heat on a range of outcomes including energy use, air quality, public health, and the environment generally. The challenge is often to attach a monetary value to these actions and potential impacts so that a judgment can be made on how to achieve the greatest benefit from a given investment. One of the earliest CBA on the UHI was performed by Rosenfeld et al. (1998), who combined detailed modelling of building energy use with a mesoscale weather model to examine the impact of introducing cooler roofs and cooler paving, as well as planting 11 m trees, on the UHI of the Los Angeles basin. The results suggested a significant lowering of the UHI, a decrease in the frequency of high-ozone events, and a substantial reduction in the air conditioning costs for homes. The authors estimated that the impacts of these initiatives were worth $500 m per year. This example illustrates the complexity of assessing the direct and indirect impacts of UHI actions.

Perhaps the simplest CBA that can be performed links the CUHI to building energy use using "degree days" to assess the heating and cooling demands at a place. Heating

(HDD) and cooling degree days (CDD) are calculated as the aggregated differences of hourly outdoor air temperature (T_a) when compared to threshold values. A plot of building energy use against air temperature at a place will reveal a characteristic relation with (i) a base value (e.g., lighting, hot water, and appliances) that does not vary through the year; (ii) rising energy use associated with the demand for space heating at cooler temperatures; and (iii) rising energy use associated with the demand for space cooling (Fig. 3.14). For example, the threshold values are 16 and 22 °C for heating and cooling, respectively, in the EU. Here, the bulk of domestic energy is used for space heating (about 125 kWh m^{-2}) and the cost of electricity is about €0.20 per kWh. To estimate the energy use that can be attributed to the urban temperature effect, the annual CUHI would be used to recalculate the HDD and CDD and associated expense (or saving). Against this value, the researcher would balance the costs associated with reducing the CUHI magnitude. Of course, the net cost/benefit will vary depending on the background climate, the shift in degree days associated with the CUHI, and the efficiency of building energy systems; for some climates where there is both a heating and cooling demand through the year, winter-time warming benefits may offset summer-time cooling costs.

In general, the costs and benefits (especially) may not be amenable to simple monetary evaluation and, where they are, the basis for calculation may be specific to that economy and society (USEPA, 2008). For example, the creation of a new green space in a high-density neighborhood with limited existing green cover can have multiple environmental (temperature moderation, hydrological control, enhanced biodiversity) and societal (mental health and well-being, air quality) benefits, many of which are not easily quantifiable. The co-benefits of heat mitigation strategies (that overlap closely with other environmental actions) is a topic of considerable interest in current research (Lenzholzer, 2015).

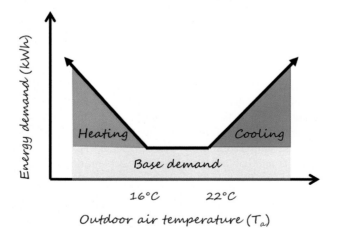

Fig. 3.14 The general relation between outdoor air temperature and the energy demand of buildings for heating and cooling (Q_F). The outdoor values (16 and 22 °C) represent the limits for heating and cooling, respectively, in mid-latitude climates.

From our perspective, it is worth placing a CBA analysis within the context of Fig. 3.1 and identifying (i) UHI risks at different scales (building, street, neighborhood, and city), (ii) who/what pays the cost of interventions, and (iii) who/what benefits. Identifying the spatial and temporal attributes of risks at the outset—analyzing the UHI in relation to the underlying physical and socio-economic attributes of neighborhoods—can provide a framework for a rudimentary cost-benefit assessment.

Keep in mind that some actions for climate management can be counterproductive, such that an action can have a positive impact on one issue (or scale) and a negative impact on another. For example, tree planting in a street will increase shading and reduce the surface UHI within the urban canopy, but if the tree canopy is extensive such that street ventilation is greatly reduced, air quality and overall comfort may be compromised. Some of the hardest questions revolve around matching climate goals at different scales, which may be associated with different processes. Many of the risks that are linked to global climate change (GCC) are driven by the urban metabolism, energy use, and the emission of greenhouse gases. There is a significant body of evidence that indicates the metabolic processes of cities are more efficient (i.e., less resource use per capita) if cities are more physically compact—higher building and population density—so that intra-urban distances and infrastructure costs are reduced. Pursuing this goal using conventional urban development policies will result in taller, closely spaced buildings with less vegetative cover (that is, more like LCZ 2 than LCZ 5, for example), which will enhance the UHI and increase the population exposed to its effects. This means that, although the energy demand per capita is reduced (positive), local environmental quality may be diminished through excessive shading, loss of ventilation, and enhanced UHI (negative). Disentangling climatic impacts at different scales to permit a useful CBA is an outstanding problem for urban climate studies.

3.7 Concluding remarks

Decisions on managing urban heat should be clear on the types of UHI, the scale and cost of potential actions, and the expected consequences. Ideally, these decisions would be based on a complete UHI climatology that would:

- describe the magnitude, timing, and spatial distribution of the temperature effect;
- establish the wider climatic background and the synoptic weather conditions that regulate its magnitude;
- place this information in the context of topographic variations and surface geometry, land cover, and land use;
- identify the neighborhoods and places where the UHI types are strongest;
- link the UHI to population exposure and vulnerability in outdoor and indoor environments.

To evaluate the efficacy of a heat mitigation/adaptation strategy, the researcher needs to gather temperature information (at the relevant scale) before and after any actions.

References

Erell, E., Pearlmutter, D., Williamson, T., 2012. Urban Microclimate: Designing the Spaces Between Buildings. Routledge.

Lenzholzer, S., 2015. Weather in the City: How Design Shapes the Urban Climate. Nai010 Uitgevers/Publishers.

Ng, E., 2009. Policies and technical guidelines for urban planning of high-density cities—air ventilation assessment (AVA) of Hong Kong. Build. Environ. 44, 1478–1488.

Rosenfeld, A.H., Akbari, H., Romm, J.J., Pomerantz, M., 1998. Cool communities: strategies for heat island mitigation and smog reduction. Energ. Buildings 28, 51–62.

Rosenzweig, C., Solecki, W.D., Romero-Lankao, P., Mehrotra, S., Dhakal, S., Ibrahim, S.A. (Eds.), 2018. Climate Change and Cities: Second Assessment Report of the Urban Climate Change Research Network. Cambridge University Press, Cambridge, UK.

USEPA, 2008. Reducing Urban Heat Islands: Compendium of Strategies. Environmental Protection Agency. Available at https://www.epa.gov/heat-islands/heat-island-compendium.

Part Two

The second part of this book establishes a UHI methodology, having now examined the scientific context for a heat island study. Our intent is to put down a set of guidelines for carrying out an observational study of the surface and canopy-level heat islands (CUHI, SUHI), and to ensure meaningful outcomes are met for appropriate audiences. In these guidelines, greater attention is devoted to the CUHI because its methods and experimental approaches are further developed than those of the SUHI.

In Chapter 4, we begin with logistical and methodological considerations that should be taken into account before going into the field to conduct a CUHI study. It is this stage of early planning that is most critical to its success, and it raises important questions about when, where, why, how, and for whom a CUHI study is to be conducted. We describe the conventional methods and instruments used to measure the CUHI, and give guidance on how these instruments should be arranged and/or transported across urban areas. This sets the sampling framework and level of experimental control for the study, allowing for certain effects—both natural and contrived—to be isolated, observed, and understood.

In Chapter 5—the second of two chapters on the CUHI—we cover the methods of data analysis and interpretation that should be employed upon returning from the field. A first, but often forgotten, step in the analysis of CUHI is the assembly of metadata to describe the field instruments, observing practices, measurement sites, and weather conditions. We carefully examine this step, along with methods of quality assurance, criteria for data stratification, and approaches to data illustration.

In Chapter 6, our focus shifts to the SUHI, but our treatment of it remains consistent with that of Chapters 4 and 5. While the aims, scales, instruments, audiences, and applications of SUHI studies are often very different from those of the CUHI, the preparatory steps (and preliminary questions) are just the same. We discuss these steps and highlight the special circumstances of a SUHI study. In doing so, it is made clear that observations of the SUHI are inherently challenged by the three-dimensional geometry of the city, the surface of which cannot be measured for its "complete" temperature, regardless of the sensing method. This constraint influences all stages of data retrieval, analysis, interpretation, and communication. We examine each of these stages, while giving emphasis to the role of satellite remote sensing.

Planning a CUHI study

4

Planning an observational study of the urban heat island entails many steps, some of which are preparatory and involve early decisions about the purpose and scope of the work. These decisions will lead you to more specific considerations about the methods of the study, the instruments to deploy in the field, and the operational definitions of the heat island magnitude. To begin, consider the first decisions as contemplative, each one allowing you to think carefully about the resources available to your study, the main objectives of the study, its intended audiences, and appropriate field strategies. These considerations should not be made in isolation of one another, as each one has a bearing on the next. Adhering to the guidelines in this chapter will help you to conduct a sound investigation of the canopy-level urban heat island (CUHI), and to produce estimates of its magnitude that meet acceptable standards for quality and reliability.

4.1 Preparatory steps

Before embarking on an observational CUHI study, it is worthwhile spending some time to ask the following questions.

4.1.1 What is the purpose of your study?

When beginning a CUHI investigation, always start with one fundamental question: What is the purpose of your study? This is not an easy or straightforward question because heat island studies are conducted for many reasons, and these can change over time and with shifts in awareness and understanding of environmental issues, both locally and globally. Technology is also changing: through history, this has given heat island studies greater detail, accuracy, and output. However, it is important for researchers to ensure that these changes do not compromise the quality of results. Traditionally, the purpose of a heat island study is to observe and monitor its climatology, meaning the long-term atmospheric effects of the heat island and its response (or statistical correlation) to measures of synoptic weather, population growth, urban expansion, and global change. A study of this purpose provides a high-level overview of the heat island without the spatial and temporal detail of its internal (intra-urban) characteristics or short-term fluctuations. A study of much greater detail will serve a different purpose, for example, to observe changes in air temperatures (T_a) in and around the city, and across hours, days, or weeks. If cleverly designed, either of these approaches can reveal the influences of land cover, surface relief, building geometry, construction materials, and soil type and wetness on T_a patterns. Furthermore,

The Urban Heat Island. https://doi.org/10.1016/B978-0-12-815017-7.00004-7

the temperature data can be used to produce isothermal maps and temperature cross-sections of the heat island and its spatial patterns.

In recent years, heat island studies have leaned away from these traditional approaches, largely because hundreds of descriptive studies from cities around the world have been documented in the published literature. Heat island patterns and processes are thus widely known and thoroughly described. In fact, one can easily anticipate the magnitude and spatial configuration of a CUHI with general knowledge of a city's geographic setting (orography, latitude, and water bodies), urban structure (building height and spacing), and prevailing weather conditions (wind and cloud). Many researchers are therefore adopting a more practical approach to heat island work, rather than an exploratory approach. Among these practical uses of a CUHI study is the assessment of heat stress in cities, mainly of pedestrians in the outdoor environment but also of building occupants in indoor settings. A CUHI study is able to provide essential data to help assess both indoor and outdoor heat stress, and to help researchers discern the effects of urban form and neighborhood design on matters of public health. CUHI researchers might therefore seek to relate their observations to the occurrence of heat illness or human mortality in the urban environment.

Global warming and rapid urbanization have motivated other applications of CUHI study, mainly to test the efficacies of heat mitigation and climate adaptation strategies in cities. These strategies involve changes to built form and human behavior that lessen the harmful or unwanted effects of urban heat. They can be deployed at a range of scales to include planting of individual trees; replacement of hard or dark surfaces with pervious or light-colored materials; and integration of passive heating and cooling systems into building designs. A CUHI study of this intent will sample T_a at sites before and after the implementation of these strategies, or synchronously at sites of similar design apart from one critical element (e.g., green cover). In both cases, attention must be given to the experimental setup of the study, and to neighborhoods that are vulnerable to extreme heat. Design elements and spatial scales applicable to city and neighborhood planning will therefore become relevant to an observational CUHI study.

4.1.2   What resources are available to you?

Important to any CUHI study is the availability of resources to carry out the study and to meet its objectives. Necessary resources include time, personnel, equipment, transportation, and finances. Taking stock of such resources allows you to set a manageable scope and outcome for your investigation. "Time" is perhaps paramount: if you have several years to carry out the study, you might consider observing the climatology of the heat island effect using official climate stations at permanent or semi-permanent sites. However, if you have only a few days or weeks, a more accessible approach might involve mobile surveys across the city, or within its individual neighborhoods. These surveys require only a few hours to complete, and can be repeated daily or nightly to acquire interesting insights on the heat island effect and its local and microscale climates. In this way, the desired purpose of your study is dependent upon the timeframe that is available to you.

Access to meteorological instruments also influences the design of a CUHI study and its achievable aims. You may have access to a science laboratory that is equipped with thermometers that can be placed for long periods of time at selected sites in an urban-rural area, or otherwise transported through the city via bicycles, automobiles, or on foot. If instruments are not available to your study, you may need to purchase or borrow a set, or else seek temperature data from existing climate or weather stations—these are often found at airports, university campuses, or other institutional settings. In many cities, urban meteorological observations are ongoing through cooperative (or volunteer) networks of stationary instruments, or through citizen-based efforts to "crowdsource" environmental data. Low-cost thermal sensors are also available on the market and can be purchased in large numbers to enable a detailed heat island study.

4.1.3 Who is your target audience?

The target audience will partly determine the purpose of your study (and vice versa). From the outset, your audience can be simplified into one or two groups: scientists and/or policy makers. The former has received much of the work on heat island studies over the past century, while the latter has become increasingly relevant in recent decades. Departments of academic study now concerned with urban climate science have expanded greatly from the traditional disciplines of geography and meteorology, to applied disciplines of building architecture, landscape ecology, urban planning, civil and environmental engineering, public health, and information science. Audiences are now easily found among these fields that can help to manage the societal implications of urban heat islands, namely through heat mitigation, climate adaptation, and ecological protection. In these cases, a CUHI study might be targeted to a group of neighborhood planners, public health officials, landscape architects, or broadcast meteorologists. Each of these groups requires a heat island investigation of appropriate design and of relevant spatial and temporal scales. It is time well spent for you to contact members of your targeted audience ahead of data collection, and to inquire about the type and quantity of data that might be helpful to their work, and how to communicate those data in ways that are meaningful and impactful.

4.2 Designing the study

After contemplating the questions raised in Section 4.1, you are much nearer to declaring a purpose and audience for your study. You can then gather the human and material resources needed for that study and its target community. Your thoughts on the foregoing questions now dictate other key considerations, such as the sampling framework (density and frequency of measurements), the means to collect field data (moving platforms or fixed stations), the type of instruments to measure T_a (thermistors, glass thermometers, low-cost sensors), the need for experimental control of nonurban influences (weather, surface relief), and the operational definition of heat island magnitude (spatial averages vs. two-point comparisons). We will discuss each of these individually, noting that one is not exclusive of the others, nor is it independent of the preparatory questions already asked in Section 4.1. All of the questions and considerations raised in this chapter are intersecting: for example, the stated purpose of your study will partly determine the resources needed, as well as the audience that is most interested in the results. The stated purpose also influences the choice of sampling framework and operational definitions, which in turn inform the selection of methods and instruments with which to gather the temperature data. These relations are equally true in reverse order—that is, the instruments available to you will influence the choice of field methods, the target audience, the study purpose, and so on. Thus one cannot, and should not, make final decisions about a planned heat island study until all factors involved in data collection, analysis, output, and application have been carefully considered.

Possibly the most fundamental consideration you will face when planning a CUHI study is the methodology used to measure T_a in the field (Fig. 4.1). There are two conventional approaches—mobile surveys and stationary surveys—and both have been used for many decades in urban climate work. However, the approaches have vastly different resource needs and data yields, while serving very different aims (Table 4.1). One must fully understand these differences before selecting the approach that is best suited to the goals and outcomes of the study.

In the following sections, we give an overview of these approaches, and then look to the many types of thermometers that can be used in a heat island study. We also discuss available frameworks to inform the placement of thermometers and their sampling frequencies, as well as experimental controls to isolate "urban" effects among the many other nonurban (and potentially confounding) effects on the heat island, and lastly the operational definitions to convey a set of procedures for measuring the heat island magnitude.

4.2.1   Methods of observation

Urban heat islands in the canopy layer are conventionally measured with ground-based methods. These involve stationary surveys with temperature sensors placed at fixed points in and around a city, or mobile surveys with sensors carried across an urban-rural area by foot, automobile, bicycle, or motor scooter. Sensors that have remained fixed in place since before urbanization (i.e., pre-urban times) are ideal for observing historical changes in heat island magnitude, but these rarely exist in most cities. Use of fixed urban-rural pairs or networks of sensors is nevertheless the norm for most stationary surveys of the CUHI. As shown in Fig. 4.1, both stationary and mobile surveys provide a simple means to measure T_a in cities of any size, location, or description. Table 4.1 summarizes the main differences between the two approaches in terms of their resource needs, data yields, primary uses, experimental controls, and notable drawbacks. These should help you to select an appropriate survey method to meet the aims and available resources of your study. In all CUHI field studies involving more than one thermometer, these should be compared against a reference

Preparation

Identify background stations Ⓐ Ⓑ

High-quality observations at sites with known exposure and source area (fetch)

Observations of
- air temperature
- wind
- cloud
- precipitation
- radiation

Sub-hourly observations

Map land use, land cover, and topography (e.g., Local Climate Zones, LCZ) to design a sampling framework

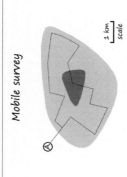

Stationary survey

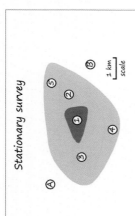

Spatial sampling to capture urban extent and variations in the urbanised landscape.

Select sites to represent LCZ types and to control for topographic effects (hills, valleys, water bodies).

Select thermometers to measure with superior resolution (± 0.1°C), certainty (± 0.2°C), and a ventilated radiation shield.

Chose precise location for thermometer and shield. Record metadata for each site and assess its exposure.

Choose logging system to record at sub-hourly rates.

Calibrate thermometers before deployment in the field. Arrange for frequent site visits.

Mobile survey

Spatial sampling to capture urban extent and variations in the urbanised landscape.

Select route to represent LCZ types and to control for topographic effects (hills, valleys, water bodies).

Design route to be completed in 1–2 h ideally, start and finish at a fixed station, forming a loop.

Select thermometers to measure with superior resolution (± 0.1°C), certainty (± 0.2°C), and response times (seconds).

Calibrate thermometers before deployment.

Choose time and weather conditions.

Record the position and speed of the mobile platform (e.g., automobile, bicycle).

Fig. 4.1 Planning a canopy-level urban heat island (CUHI) study.

Table 4.1 Summary of differences between stationary and mobile CUHI surveys.

	Stationary surveys	Mobile surveys
Primary uses	To observe (i) the climatology of heat islands; (ii) the combined and/or separated effects of heat islands and global warming; (iii) the seasonal and/or annual temperature conditions inside LCZs; (iv) the heat island magnitude over time and with environmental change.	To observe (i) the local and microscale effects on heat island magnitude and morphology; (ii) the statistical relations between heat island magnitude and urban geometry, land cover, and daily weather (wind, cloud); (iii) the microscale temperature variations inside LCZs.
Resource needs	Climate stations and/or temperature sensors at fixed locations in and around the city. Minimum of two locations (one urban, one rural), preferably a network of multiple urban and rural stations.	Mobile platforms (e.g., automobiles, bicycles). One or more temperature sensors, data loggers, and GPS units. Field assistants to help with data collection and transport of equipment.
Duration of field campaigns	Usually long—several seasons or years.	Usually short—several days or months.
Temporal resolution of survey data	Usually high—sampling frequency may be continuous or short time-intervals (seconds, minutes) throughout the campaign.	Usually low—sampling periods typically last a few hours and are infrequently repeated during the campaign.
Spatial resolution of survey data	Usually low—depending on the number of stations or sensors in the study area (i.e., sampling density).	Usually high—depending on the number of measurements taken along the survey route and selection of sites. If measurements are made at short time-intervals (seconds), not all sites will be locally representative.
Locally representative sites	Difficult to achieve with pre-existing climate stations or temperature sensors, especially if sited for purposes other than heat island observation. Possible with careful siting of fixed sensors.	Easy to achieve with careful design of survey routes and selection of sites. If measurements are made at short time-intervals (seconds), not all sites will be locally representative.
Experimental control	Control of surface effects (relief, urban form) is difficult due to low (and often pre-determined) spatial resolution of the measurement sites. Control of time (synchronous measurement) and weather effects (wind, cloud) is simpler due to long duration of field campaigns, large sample sizes (number of repeated observations), and automated measurements.	Control of surface effects (relief, urban form) and weather effects (wind, cloud) is simple due to high spatial resolution of measurement sites, and opportunity to sample where/when desired (range of times, sky conditions). Control of time (synchronous measurement) is difficult due to low temporal resolution of sampling periods, the need for time-temperature correction, and small sample sizes (number of repeated measurements).
Notable drawbacks	Difficult to find safe, secure, accessible, and locally representative sites. If pre-existing stations or sensor networks are used, these are often of low spatial resolution or without site metadata. Volunteer, amateur, and crowd-sourced networks increase the quantity of data, but reduce the quality.	Survey routes are usually restricted to public roads. Roads may not be representative of the local area through which they pass. In large cities, it is difficult to cover the entire urban area in a reasonable time period. Late-night data collection required to observe the maximum heat island magnitude.

thermometer at the start and end of the study to ensure accuracy and compatibility, particularly if the technical or material characteristics of the thermometers differ (e.g., shielding, time-constants). If the study period extends for several years, regular inspection and comparison of thermometers is needed to identify drifts in accuracy and aging of hardware components.

4.2.1.1 Stationary surveys

Stationary surveys consist of temperature sensors fixed at one or more centrally located urban sites, which are normally paired with one or more rural (or background) sites (Fig. 4.1). The sensors may already exist at an established climate station or meteorological observatory (e.g., airports or universities), or they may require installation at chosen points throughout the city and its environs. If the network of sensors or stations is sufficiently dense, it can highlight certain features of the landscape, to include the thermal influences of land cover, land use, urban form, and human activity. Stationary surveys are favored for monitoring the temporal rather than spatial characteristics of the CUHI (Table 4.1). If the sensors are fixed for many months or years, the output can help researchers to study the climatology of heat islands, observing their changing magnitudes with weather conditions, specifically wind, cloud, humidity, and atmospheric stability, and with long-term population growth and urban expansion. Stationary surveys also allow for calculation of seasonal and annual means in CUHI magnitude and their trends and fluctuations over time. If dozens of sensors are positioned in the urban area, this can reveal the internal (local) temperature patterns of the heat island. The difficulty of finding secure sites at which to place the sensors is a drawback of the stationary approach, as is the limited spatial resolution that follows from a single pair or reduced number of sensors or observing stations, especially in cities and/or metropolitan regions with expansive and heterogeneous forms.

A stationary survey may involve historical networks of "official" climate stations that have operated for many years, and that are maintained by government or other institutional bodies to meet WMO standards for meteorological observations. Alternatively, the survey may involve a comparatively crude arrangement of low-cost sensors or amateur weather stations of "unofficial" status. These can be found in most cities and are often run by volunteers, casual observers, or citizens with lay interests in weather and climate (e.g., gardeners, farmers, school students, nature enthusiasts). Sites for unofficial networks are typically located in school grounds, work yards, residential yards, community gardens, university campuses, or at utility and transport facilities. If the exact locations of the sensors are known, one might visit the site to assess its instrumentation and surrounding landscape; if this is not possible, aerial or satellite imagery can be obtained from a geobrowser (e.g., Google Earth) and used for the same purpose.

Other networks have been established through crowdsourcing of temperature data. This involves untrained observers (i.e., the public) installing low-cost outdoor thermometers at places of convenience (e.g., a window sill or building wall of a private residence), with the data automatically uploaded to the Internet to provide real-time information. The data can be used to expand the temporal and spatial coverage of existing weather and climate observations in the area. Crowdsourcing may provide an accessible, cost-efficient means of monitoring and mapping T_a patterns in cities. The sensors are easy to install and maintain, and the data are usually available to the public. However, despite the obvious benefits of crowdsourcing, the data have numerous drawbacks for urban climate studies. First, they are often not accompanied by supporting information (i.e., metadata), which means the public knowns little or nothing about the exposure or mounting conditions of the temperature sensors. Second, the sensors are often installed in peculiar places by observers with no intent (or no trained experience) to measure urban temperature effects, and with no knowledge of WMO guidelines for siting meteorological instruments in urban areas. Third, the sensors are not equipped with electric fans or radiation shields, nor are they calibrated regularly to a common reference. Fourth, the distribution of the sensors is biased to populous areas of the city, and to higher-income neighborhoods. This means that gaps in coverage will likely co-exist with parks, business districts, industrial zones, and low-income communities. Crowd data are therefore generally ascribed poor or unknown quality, and a clear role for their use in routine urban climate studies has yet to be demonstrated. However, the development of quality assessment procedures for crowd data is a promising area of work that might eventually allow their use in formal CUHI studies (e.g., Meier et al., 2017). This potential will be greatest in mid-latitude cities where crowd-based sensors exist in large numbers (e.g., Europe), and least in lower-income regions where the sensors are few in number. Nevertheless, the use of crowd data in any CUHI study must be treated with seriousness—considerable effort is needed to assess and retain only those data of highest quality and known spatial validity.

Stationary surveys sometimes involve sensor networks used for an alternative, albeit related, purpose to CUHI observation. Such networks might exist to monitor environmental elements other than (but often including) air temperature, such as carbon dioxide, particulate matter, precipitation, or stream flow. These networks can be useful to a CUHI investigation, but one must always be cautious of this approach because the systems were not originally designed for a heat island study. The instruments might therefore be sited in places that are inappropriate for sampling local air temperatures, or that do not satisfy relevant WMO guidelines. It is essential to assess the suitability of these sites and their instruments *before* using them. This makes the reporting of site metadata paramount to a study's integrity, particularly if the instruments are sited according to internationally agreed standards in other areas of work, such as aviation, agriculture, hydrology, or air pollution. In recent years, purpose-built networks for urban climate monitoring have been established in some cities. These systems are preferred for CUHI studies because the stations should adhere to WMO guidelines for urban observations (e.g., Aguilar et al., 2003; Oke, 2006). However, the networks are expensive to maintain and difficult to configure due to restrictions on site access, security, and safety. Thus the lifespan of the systems is often quite short.

4.2.1.2 Mobile surveys

An alternative approach is a mobile survey using road vehicles equipped with temperature sensors. Survey routes through urban-rural areas may be linear or circuitous, and they may be designed to measure a low- or high-density distribution of temperatures in a relatively short period of time (Fig. 4.1). Automobiles are the favored mode of transportation during a mobile survey, although bicycles, motor scooters, and foot traverses are also used. Temperature sensors are normally attached above the roof or bumper of a car, such that engine or exhaust heat is avoided by the sensor while the car traverses the city. Mobile surveys are suited for observing microclimatic featuresof an urban environment, such as a park or a street canyon, or for investigating the effects of neighborhood or block design on local temperature regimes (Table 4.1). The data gathered from these surveys are used to produce isothermal maps or temperature cross-sections of the (horizontal) temperature field, which reveals the outer boundary of the heat island that normally coincides with the edge of the built-up area (Figs. 1.4, 5.4, and 5.6). Within the built-up area are temperature peaks associated with the city core and/or compact subcenters, with spikes and dips that reflect sudden changes in building density and land cover. Mobile surveys can also be used to measure temperatures under a range of synoptic weather conditions and at various times of day. Single-vehicle surveys are uniquely advantaged because errors from instrument calibration between two or more vehicles moving through the study area are eliminated; a fleet of moving vehicles, on the other hand, can cover a much larger area and obtain larger datasets than a single vehicle. This is particularly helpful for field studies in large metropolitan regions. Finally, despite the high spatial resolution of mobile sampling, its temporal resolution is limited as sampling periods rarely exceed a few hours.

4.2.2 Instrumentation

The standard glass thermometer is the oldest instrument employed by CUHI investigators: it has been used for more than two centuries, dating back to the classic heat island studies of the 1800s. However, in recent years the use of automated, lightweight, and low-cost electrical sensors has become increasingly popular. The choice of instruments to be deployed will depend mostly on practical matters (e.g., cost, accuracy, portability), but also on theoretical concerns relating to the scope and purpose of your study. There are important guidelines to follow for all instruments used in CUHI work, regarding their security, shielding, ventilation, and technical performance. These guidelines are provided in the following sections, along with an overview of main instrument types.

4.2.2.1 Instrument type and technical performance

Liquid-in-glass thermometers

The earliest systematic studies of the heat island effect were conducted with liquid-in-glass thermometers containing mercury or alcohol. Liquid that is encased in a capillary tube expands and contracts as air temperature rises and falls, indicating temperature values against scale markings that are engraved on the tube or its backing. These instruments were adequate for a basic study of the CUHI in the 19th and 20th centuries, and are certainly adequate today. Glass thermometers are reliable, inexpensive, simple to use, and easily installed (or transported) in urban and rural areas. Psychrometers are a more sophisticated type of glass thermometer: they measure dry- and wet-bulb air temperatures (and thus relative humidity), and require users to manually whirl or forcefully ventilate the instrument until its bulbs reach thermal equilibrium with the ambient air. Assman and sling-type psychrometers are generally used by science students and field researchers, rather than by weather hobbyists or lay observers.

For stationary surveys of the CUHI, glass thermometers can be placed at locations in the study area that allow observers to read and record T_a at selected times of day, usually for a period of several months, seasons, or years. This approach can be time consuming and it requires a lengthy commitment from the workers (or volunteers) involved in a CUHI study. For mobile surveys, glass thermometers can be fixed to a bicycle or automobile, or carried by hand while traversing on foot across an urban area or through selected neighborhoods. If using a hand-held thermometer, one should be very careful not to grip the instrument near its base where the spherical bulb is located, while holding it at arm's length upwind of the body (to avoid passage of heat from the body to the thermometer). One must also be aware of the response time (or "time-constant") of a thermometer used in a CUHI study. Response times indicate the interval of time (on the order of seconds to minutes) needed for a thermometer to fully respond to a change in T_a. The interval begins when the thermometer is subjected to a temperature change, and ends when it reaches a steady recording of that change. Due to the slow response of most glass thermometers, one needs to interrupt the traverse at each measurement point and record the temperature only after the instrument has adjusted to the ambient air. This procedure can take several minutes, adding considerable time to the duration of a survey, especially if there are tens or hundreds of points to visit. If temperatures are recorded before the thermometer reaches equilibrium with the ambient air—that is, by not allowing the required interval of time to pass while the instrument is held stationary—the measured values will not coincide with air conditions at the instant (or position) of their measurement. This leads to lag errors that might invalidate any observed relations between T_a and the measurement environment. The severity of these errors will depend partly on the sampling framework of the temperature survey (Section 4.2.3).

Also relevant are the measurement resolution and uncertainty of standard glass thermometers. Measurement resolution—which is often called "precision"—is the smallest change in temperature that can be detected by the thermometer; in most cases it is 0.2–0.5 °C. Measurement uncertainty expresses the "accuracy" of the thermometer in quantitative terms—that is, the interval of values within which the true temperature lies. "Accuracy" is strictly a qualitative concept that describes the closeness of agreement between a measured temperature value and the true value, and should not be confused with precision (WMO, 2008). For most glass thermometers, the uncertainty is ± 1–2 °C. These specifications may not be suitable for temperature surveys aimed at detecting small-scale fluctuations in the CUHI, associated with variations in land cover, building geometry, or human activity. For meteorological studies at city block or neighborhood scales, the measurement error of a glass thermometer is often greater than the observed inter-site temperature differences. Moreover, glass thermometers cannot automatically record T_a, and manual reading of the instrument at minute or hourly intervals over the long term is not practical. For these reasons, glass thermometers are rarely used in contemporary heat island studies.

However, researchers using historical temperature data from decades or centuries ago, or having interest in the classic work of the historical period, will need to understand the operations and specifications of standard glass thermometers (Stewart, 2019).

Thermistors, thermocouples, and electrical resistance thermometers

As the aims of heat island investigation have evolved, so too has the instrumentation. This has been advantageous to researchers wishing to conduct temperature surveys with greater spatial and temporal detail. Glass thermometers have therefore been superseded by automated, fast-response electrical thermistors, thermocouples, and resistors, many of which can measure T_a with superior resolution (0.1 °C), certainty (± 0.2 °C), and response time (on the order of seconds). Thermometers operating by the electrical resistance of metal types are capable of remote reading and recording of temperature data, and respond quickly to temperature change across short time periods and distances. This is because the thermometers have thin and small elements with negligible heat capacity. The instruments can be rigged to data-recording systems that log temperatures automatically at high frequency (minutes or seconds), or even continuously. Such systems can automatically transmit data to the Internet or to a central processing system (via communication and data transmission networks) for real-time monitoring of T_a. This arrangement is often preferred for CUHI studies (whether stationary or mobile) because the time needed to record, gather, and later analyze the temperature data is greatly reduced. During mobile surveys, quick-response sensors are capable of discriminating subtle changes in temperature at much-reduced spatial and temporal scales. The sensors are compact and lightweight for easy attachment to moving automobiles or bicycles, or to stationary fixtures such as lamp posts or traffic signals. High-end models of these instruments can be very expensive, but costs have dropped considerably in recent years.

Low-cost sensors

The introduction of low-cost sensors has aided heat island research worldwide, especially in lower-income countries. Low-cost sensors are now available to researchers who could not otherwise afford meteorological instruments for field studies. A popular brand of low-cost thermometer is the *iButton* logger, whose size is no larger than the thumb of your hand. It can therefore be placed securely and inconspicuously at urban or rural sites. *iButtons* can be configured for automatic upload to the Internet, or for internal storage of data for manual transfer by field workers. *Netatmo* weather stations are configured in a similar way and are easily installed in outdoor settings. However, the main drawback of these sensors is their reduced measurement resolution (0.1–0.5 °C), reduced certainty (± 0.5–1 °C), and lack of radiation shielding.

A second, and very different, example of a low-cost sensor is the battery thermometer contained in smart/mobile phones. Given the ubiquity of these devices in cities, some researchers argue that this presents an opportunity to retrieve many thousands of geo-located crowdsourced temperatures that can be used to estimate the air temperature, and subsequently to improve the spatial and temporal resolution of heat island mapping. However, the benefits of this approach to urban climate science are largely unproven, and adherence to established guidelines for heat island observation—with respect to site selection, sampling frameworks, and provision of metadata—is virtually impossible with personal mobile phones. If one chooses to use such data in a CUHI study, this must be done with full awareness for the location, accuracy, and exposure of all measurements.

4.2.2.2   Instrument shielding

After deciding on an instrument type that is suitable and affordable to you, consider its mounting, shielding, and ventilation when installed in the field. This is to make certain that the *measured* T_a are closely adjusted to the *actual* T_a. Note, however, that the actual temperature can never be obtained with instruments because the instrument itself perturbs airflow and radiation exchange, and its response to temperature change is always lagged. Sensors of tiny construction will cause less perturbation and have shorter lag times to the measurement environment than larger instruments, such as the standard glass thermometer. To minimize errors from atmospheric radiation, wind, precipitation, and pollutants, thermal sensors require shielding and ventilation that adheres with WMO (2008) guidelines. These guidelines ensure that sensors are housed in standard weather shields (also called screens, shelters, or boxes) constructed from materials of low thermal absorption (e.g., wood, plastic, polished metal) and high solar reflectivity (white in color) to reduce conduction of heat toward the interior of the shield (Fig. 4.2). The shield must also allow passage of air through its interior and over the sensor, while protecting the sensor from direct solar radiation and contact with rain, fog, snow, and airborne dust, dirt, and pollutants. In short, a well-designed shield ensures that the temperature of the interior air is nearly equal to that of the exterior.

A widely recognized shield called the "Stevenson screen" has been used for more than a century to provide shelter for ground-based meteorological instruments. The screens are white wooden or plastic boxes of low heat capacity and with louvered sides, positioned firmly atop a sturdy support-stand at 1.25–2 m height (Fig. 4.2). The screens help to standardize air temperature measurements and to mitigate the variable effects of radiation from nearby trees, buildings, ground cover, and other surface features. Most official observing stations now use Stevenson-type screens for housing thermometers (Fig. 4.3). Smaller gill-type varieties of the Stevenson screen are also used, with multiple light-colored plastic or metal discs layered vertically around the sensor. Makeshift screens can be constructed with common household materials if the investigators wish to use their own instruments (rather than official stations) and to site the instruments at places of their own choice (Fig. 4.2). In this case, the general principles of solar reflectivity, thermal conductivity, weather resistance, and free exchange of air must be considered in the screen design. In other cases, thermal sensors may already be enclosed in a radiation shield when purchased from a manufacturer or retailer, and may not need further protection when installed in the field. For a mobile survey, radiation shielding is necessary only for daytime runs; at night, solar receipt at the sensor is zero. However, protective shielding is still advised to keep the sensor clean and dry during a survey at any hour.

Fig. 4.2 Styles of instrument shielding for temperature sensors in CHUI surveys. (A & B) Standard Stevenson screens (or "weather boxes") of wood or plastic construction; (C & D) gill-type screens with layered metal or plastic discs around the sensor; (E & F) makeshift screens constructed of common household materials.
From I.D. Stewart, except for image (F) from R. Kotharkar.

Fig. 4.3 Interior of a Stevenson screen. Shown are three pairs of instruments to measure air temperature and relative humidity: (A) dry and wet bulb glass thermometers; (B) maximum and minimum glass thermometers; and (C) dry and wet bulb electrical resistance thermometers. The Stevenson screen is part of a standard weather station at the Darwin International Airport (Australia).
From Wikimedia Commons, https://creativecommons.org/licenses/by-sa/3.0/au/deed.en.

4.2.2.3 Instrument ventilation

We have so far only hinted at the importance of ventilation for instrument shielding. Having louvered sides on the shield allows for natural (i.e., unforced) mixing of interior and exterior air around the sensor (Fig. 4.3). Natural mixing is acceptable if the shield is placed at sites that are exposed to the movement of air across the measurement area, so that the temperature inside the shield is representative of the outside area to be monitored. If the shield is placed at a site with greatly restricted or perturbed airflow (e.g., next to a building wall), ventilation should be forced through the enclosure by an electric fan that draws air at an even pace into and out of the shield. In many heat island studies, forced ventilation may not be required because sites in urban and rural settings are normally well exposed to airflow. However, if the natural wind speed is very low ($< 1\,\mathrm{m\,s^{-1}}$) and forced ventilation is not provided, the temperature inside the shield may not represent the outside air and biases will arise in the instrument readings. For mobile surveys, ventilation is made sufficient by the movement of the vehicle, except when it stops (e.g., at street intersections or in slow-moving traffic) or moves too quickly (causing dynamic heating of the sensor). For this reason, forced ventilation is still preferred because it allows for continuous and steady air exchange with the sensor, and it reduces measurement errors from the emission of vehicle engine and exhaust heat.

4.2.2.4   Instrument mounting

It is difficult to set rules for where and how to mount sensors in urban environments—these will vary with the goals of the investigation and with site access and instrument security. However, there are two key considerations for mounting instruments in a CUHI study: (i) height placement of the sensor above ground, and (ii) the support structure from which the sensor and its shield are mounted. These considerations do not normally apply to sensors already housed in a standard Stevenson screen or official weather box, which should comply with WMO guidelines for meteorological observations. Sometimes, however, Stevenson screens are located in unconventional places, such as building rooftops. WMO (2008) guidelines state that thermometers should be placed 1.25–2 m above ground, preferably cropped grass, and away from the immediate effects of trees, buildings, and other surface objects. These guidelines certainly apply to rural areas, but less so to urban areas where the observed effects of trees and buildings are deliberately sought by urban climatologists. Oke (2006) therefore adapted the WMO guidelines for use in urban areas, to allow height placements to be relaxed, i.e., higher than 2 m above ground (but never lower than 1.25 m where the vertical temperature gradient is large). In urban areas the air is well mixed, and thus air temperatures above 2 m will not exhibit sharp vertical gradients. This is certainly the case for street canyons experiencing cross-flow and channeling of winds (Fig. 2.18). Nocturnal air temperatures measured at 5–10 m above ground are therefore unlikely to differ greatly from those measured at 2 m above ground. Furthermore, in urban areas, it may be necessary to place sensors well above standard screen height to avoid theft or vandalism. However, placement of instruments on rooftops (above the urban canopy layer) should be avoided because the radiation and energy balances there are very different from those at ground level (below the mean canopy height). This is a crucial point because rooftop sites are often used together with ground-level sites in CUHI studies. If rooftops are unavoidable (e.g., for security reasons), this can be justified but it must be disclosed in the heat island report along with metadata to convey the location, instrument exposure, and mounting conditions of the site.

Support structures from which to mount temperature sensors for stationary and mobile surveys of the CUHI are illustrated in Fig. 4.4. For stationary surveys, these include lamp posts, signposts, fence posts, utility poles, traffic signals, building walls,

Fig. 4.4 Common support structures for mounting temperature sensors in stationary or mobile surveys of the CUHI. Arrows indicate position of sensor. (A) Stevenson screen on podium rooftop; (B) lattice tower on building rooftop; (C) traffic light-signal; (D) lattice tower on ground; (E) wall bracket in street canyon; (F) street lamp post; (G) steel post on residential lawn; (H) utility box on lawn; (I) power pole; (J) automobile rooftop; (K) bicycle steering post; (L) arm of a field worker.
From I.D. Stewart, J. Unger, J. Basara, Y. Huang, and G. Guzmán.

and so on. These are ideal if they do not greatly perturb the airflow and radiation balance of the measurement site. Attaching sensors to tree trunks (i.e., below the crown) should be avoided, unless it a tree's microenvironment that is the subject of study (e.g., for outdoor thermal comfort). Sensors can be mounted to building walls of street canyons if the mounting bracket extends outward from the wall by at least 1 m, to avoid radiative heating of the sensor. This arrangement is acceptable for a study of the street canyon climate. For a mobile survey, sensors are normally mounted to the body of a vehicle, e.g., its door, rooftop, or front/rear bumper (Fig. 4.4). To be consistent with the height placement of stationary sensors, mobile sensors should be mounted at 1.25–2 m above ground. Of more concern, however, is the position of the sensor relative to the vehicle engine and exhaust pipe. Ideally, the sensor should be upwind or otherwise away from the flow of engine and exhaust heat. The same principle applies to surveys conducted on foot or bicycle: mount the sensors 1.25–2 m above ground and upwind of the observer's body to avoid the influences of metabolic heat (Q_M) on the temperature readings. For a bicycle survey, this means extending the sensor forward or sideways from the steering post; for a foot survey, it means holding the sensor at arm's length upwind of the body. If winds are light or calm ($< 1\,m\,s^{-1}$) during a foot survey, the observer should manually ventilate the sensor by gently swaying it from side to side at an even pace.

4.2.3   The sampling framework

Three aspects of the sampling framework for CUHI studies will be covered in this section: (i) selection of representative sites; (ii) sampling density and frequency of temperature measurements; and (iii) experimental control of interfering effects. These aspects are relevant to both stationary and mobile surveys, and their treatment is dictated by the purpose of a CUHI study.

4.2.3.1   Selection of representative sites

The notion of *representativeness* is defined by the WMO (2008) as *the degree to which an observation accurately describes the value of the variable needed for a specific purpose*. For the purpose of a canopy-layer UHI study, the variable needed is air temperature and it should be valid beyond the immediate position of the sensor to a few hundred meters in the horizontal scale. If the air temperature is known to change appreciably with increasing distance from the sensor, the measurement site might not be representative of the local scale. To phrase this differently, imagine moving the sensor 100 m (or less) in several directions from the site: if the topography (i.e., surface relief, land cover, urban form, or LCZ class) changes abruptly with the move, the site is probably not suitable for a CUHI study at the local scale. Examples of representative and nonrepresentative siting are given in Fig. 4.5. Sundborg (1951) was the first heat island investigator to consider this issue in great depth. He developed statistical approaches to identify locally reliable sites, giving careful thought to the perils of nonrepresentative siting. Sundborg defined "representativeness" by the statistical deviation of a site's temperature value from the observed mean value of the local area surrounding the site. Low deviation meant that the measurement site and its vicinity were characteristic of the local area, due to a reasonable degree of uniformity in land cover, surface roughness, and sky exposure around the site. Sundborg's treatment of representative siting remains central to any heat island study today.

The theoretical framework within which to judge site representativeness is based on the notion of turbulent "source area" (also called "footprint" or "circle

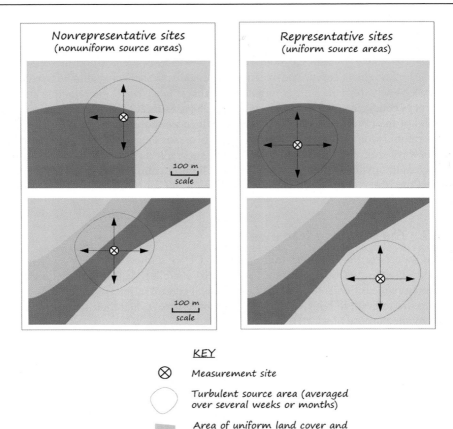

Fig. 4.5 Selection of representative measurement sites for CUHI observation. At left, the sites are not representative of the local scale because they are situated next to an LCZ border (i.e., a sharp change in land cover or surface geometry). In this case, source areas overlap with two or more LCZ classes. The sites at right, however, are locally representative because they are situated away from LCZ borders and their source areas lie within one LCZ class.

of influence") for an exposed thermometer. This is the area from which the surface air and its properties are carried up and over (by turbulence) to the sensor. The area grows (diminishes) with increasing (decreasing) atmospheric stability and simple (complex) surface geometries. As a rule of thumb, it extends horizontally from tens to hundreds of meters around and upwind of the sensor. Averaged over long periods of time (weeks to months) and many wind directions, the source area becomes approximately circular in form, with the sensor located at the center point (Fig. 4.5). It is therefore possible that, for a short period of time (a few hours), a thermometer is representative of the local scale if wind direction is constant and the source area is contained to a single LCZ class. However, if wind direction and source areas change over time to include more than one LCZ class, the sensor may not be representative of the local scale. CUHI investigators should bear these relations in mind when selecting sites, assessing spatial validity, and aligning source areas with study aims and applications.

The extent to which the measurement sites are judged to be representative for a heat island study depends fundamentally on the specific goals (and spatial scales) of the study. If the main objective is to study heat island climatology at local or city scale, a representative site must describe an area that meets this goal. This could be a street canyon in the central core of the city, in which case the canyon must characterize the area for its geometry (street width, building height), materials (surface cover, building construction), and human activity (vehicle flow, building energy use). In parts of the city that are less built-up (e.g., open-set houses), representative sites will take a different form to include more vegetation, fewer and smaller buildings, and reduced traffic flow. Temperature observations to represent these areas should be taken at appropriate locations and measurement heights, such that the source areas of the sensors overlap with a length scale that is consistent with the study objectives, and that captures the surface character and human activity at the same scale. On the other hand, these same sites might make poor choices for a study that aims to assess the outdoor thermal comfort of pedestrians. In this case, temperatures measured at places where people congregate—e.g., an urban park, a street-side bus stop, a pedestrian square, or a traffic intersection—are all good choices, even though the temperatures might represent microenvironments that extend only a few meters from the measurement site.

Aligning site locations with study aims applies to both stationary and mobile surveys. This is particularly true of long-established climate stations: their sites are fixed and cannot be moved for alternative uses or modified source areas. If the sites are not suitable for a heat island study, the investigator can (a) use the sites anyway, but with full explanation as to why they are not suitable and what effect this might have on the results; or (b) use other sites at more representative locations. One is often faced with this decision when using climate stations at airports, which have historical records that can be advantageous to CUHI studies, but whose locations are seldom representative of either the urban or rural environment. This is also a problem for volunteer or crowd networks of sensors, which can be exceedingly large (involving many hundreds of sites) and thus individual assessment of site representativeness is not possible. Moreover, the sites are not usually accompanied by metadata, meaning the exposure, height placement, and, most critically, source area of the sensors are unknown and cannot be included in a heat island report. One must therefore be wary about these sites when considering their use in a CUHI study.

The selection of sites for mobile surveys should meet similar expectations of spatial representativeness, and again these must align with the aims and scales of the investigation. The added difficulty with mobile surveys is that many more sites are used, as compared with a stationary survey. Seeking assurances that each of these sites is representative of a desired scale can be logistically demanding. Mobile data loggers, for example, can record temperatures at short time-intervals (seconds) along a survey route, such that hundreds or thousands of samples can be obtained during a single survey. For high-resolution mapping, this may be desirable and all points may be needed. For other applications, one needs to process and filter the data after collection to identify sites of appropriate representativeness. Moreover, considering that automobile surveys are confined to public roads and streets, the sites may not be representative of the land cover and surface geometry beyond the immediate area. This can lead to "confusion of scales," in which case the *microscale* setting of the measurement site is considerably different than the *local-scale* surroundings. Sites that are perhaps more representative of the local area may not be accessible by automobile or bicycle (e.g., private lawns, fenced-in areas), and thus cannot be used in the study. One needs to acknowledge this problem when it occurs.

4.2.3.2 Sampling density and frequency

An important part of site selection is the number of sites needed, and where those sites are to be placed. We cannot give rules to satisfy all purposes of CUHI study, but we do offer general guidelines for designing a simple arrangement of sites, which naturally will not apply to existing stations that are fixed and unable to be moved. A guiding principle of any meteorological investigation is that sampling density and frequency align with the time and space scales of the phenomenon being observed, in this case the heat island effect and the local climates that amalgamate that effect. The first recommendation is therefore to consider the use of a site classification scheme, such as the Local Climate Zones (LCZs), in your study area (Section 2.3.2; Tables 2.3 and 2.4). Methods to classify measurement sites (i.e., their turbulent source areas) and urban neighborhoods into LCZs are well described in the published literature and will not be dealt with here, except to say that both manual (field-based) and automated (machine-based) approaches can be used, requiring varying amounts of time, data, and technology.

There are advantages and disadvantages with any approach to LCZ classification, but the output in all cases provides a basic framework to plan a heat island study and configure a suitable number and distribution of measurement points (Fig. 4.1). Manual methods to classify areas into LCZ types require little or no technology or instrumentation, and instead use noninstrumented (visual) observations of the study area by the workers themselves. If done carefully, this can lead to reasonable estimates of a city's LCZ composition. Moreover, the manual approach normally requires users to go into the field and assess for themselves the physical character of a city or neighborhood, making skilled observations of the place. The output of this work will be a map that suggests the *actual* distribution of local climates existing in a city. Alternative methods employ aerial or satellite images of the study area, and digital datasets for land use, land cover, and building characteristics that can be combined using Geographic Information Systems (GIS) software. These approaches are fast and efficient for generating large datasets of information on urban form and function for any city, and for automated mapping of extensive areas. Quite often the desired product of this automated approach is a detailed map showing the layout of LCZ classes for a whole city, but the drawback may be that these do not depict actual climate zones at the desired local scale, owing to the nature of the geographic data used in the classification process.

Once the spatial framework (e.g., LCZ) has been established, it can be used to select measurement points and install stationary sensors at places that allow you to observe the diversity of local climates within an urban heat island (Fig. 4.1). If several sensors are available, these could be assigned to each of the LCZ classes that are represented in the study area; if only two sensors are available, these could be placed separately into one rural (or background) LCZ class (e.g., LCZ A or D) and one urban class (e.g., LCZ 1 or 2). Regardless of the sampling density, if the aim is to measure the maximum magnitude of the CUHI or its local temperature differences, the sensors should remain fixed for a period of several months or more. It can therefore be said that sampling density for stationary surveys varies with (i) the purpose and scale of the study (micro, local, regional), (ii) the availability of instruments and equipment, and (iii) the accessibility and security of sites. For mobile surveys, the study design also benefits from the use of an LCZ framework: the survey route may be configured to pass through a desired number or variety of LCZ classes in the study area, while temperatures are measured at one or many sites within each class (Fig. 4.1). Again, the sampling density will vary with the purpose and scale of the study, its logistical and technological constraints (e.g., speed of the traverse, time-constant of the sensor), and site accessibility (limited when using automobiles or bicycles).

Related to sampling density is sampling frequency. Since most instruments that are employed in stationary CUHI studies are programmed to record data at short sampling intervals (minutes, hours), the sampling frequency may already be set and sufficiently flexible to meet the goals of your study. If the stations operate for many months or years with high-frequency sampling, the dataset will be large and will require reduction (through filtering and stratification) to discern the effects of time, season, and weather on CUHI magnitude and spatial form. If, however, standard glass thermometers (e.g., maximum-minimum) are used, the sampling frequency will be low (once or twice per day). This may be sufficient to quantify CUHI magnitude at times of known maximum values (e.g., between sunset and sunrise), but if one wishes to study changes in heat island magnitude with time of day, or with cloud and wind conditions, the sampling interval will need to be shorter (e.g., hourly). Likewise, during a mobile survey, the sampling frequency can be adjusted according to the purpose of the study, the response time of the instruments, and the speed of the moving platform, or it may be set to measure temperatures continuously, in which case smoothing or averaging of the data will be necessary to obtain representative observations. However, for a mobile survey it is important that fast-response sensors are used so that the temperature measurements and their locations are in close time-space agreement. Otherwise, temperature measurements from slow-response sensors may not reflect small-scale changes in the landscape that are important to the study. This can occur, for example, with mobile surveys passing through LCZs of small horizontal dimensions (recognizing that the minimum LCZ diameter should be no less than 400–500 m). If the platform moves quickly but the instrument responds slowly, thermal changes associated with small LCZs may not be detected in the temperature record. This problem is illustrated in Fig. 4.6.

High-frequency sampling may be desired if microscale features of the CUHI are the subject of investigation. Alternatively, sampling may occur at regular time-space

Vehicle is moving at a steady speed of 35 km hr^{-1} (about 10 m s^{-1}) while sampling air temperature at predetermined sites in LCZ classes. Thermometer response time is 60 seconds. Distance between the site of measurement and the point of adjustment to the ambient air is therefore 600 m. LCZ classes with horizontal dimensions less than 600 m (e.g., LCZ B, below) will not be sampled during the survey unless the vehicle is stopped or its speed reduced.

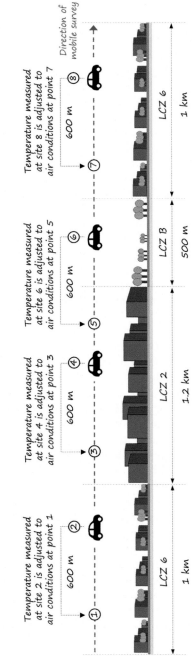

Fig. 4.6 The relation between thermometer response time and sampling framework for a mobile CUHI survey through urban neighborhoods.

intervals along the traverse route, e.g., every minute or every 100 m. This is less favorable because some of the temperatures may then correspond to sites of low spatial representativeness due to uncharacteristic surface features along the route (e.g., a tunnel, river crossing, or traffic intersection). If necessary, these temperatures can be removed during data processing. If one wants to use only locally representative sites along the route, these will probably occur at irregular intervals and should be identified before field work begins, helped by reconnaissance surveys and cartographic or photographic interpretation of the study area, as well as LCZ classification of the proposed traverse route. Despite the capacity of most data-logger systems to sample at high frequencies, one can still obtain a reasonably accurate picture of the CUHI (and its local climates) with relatively few—but strategically placed—temperature measurements in and around a city.

4.2.3.3 Experimental control

Experimental control is needed in CUHI studies to help manage the physical complexities of the urban-atmosphere system. Through control, one can reduce or remove unwanted effects that interfere with observations of CUHI magnitude, such as weather, surface relief, and water bodies. However, these effects cannot be *actively* controlled, and so *passive* approaches are necessary. If no control of observations is taken, there is risk of confounding real heat islands caused by urban effects with fictitious ones caused by other effects, such as precipitation, air mass advection, cold air drainage, or land-sea breezes (Stewart, 2011). Passive control of weather is possible through sampling designs that avoid data retrieval during frontal or unsettled weather (rainfall, strong winds). This is important because a weather front passing through an urban area may bring warm (or cold) air to one part of the city, while other parts are relatively cool (warm) (Szymanowski, 2005). If temperatures are measured synchronously in each of these parts, the observed differences are easily confused with an urban heat island effect. The same confusion arises with precipitation events. Passive control over weather effects is also possible through pre-processing of field data, such that all temperatures measured during frontal, nonstationary, or unsettled weather are excluded from analysis. Similarly, if one wishes to study the variable effects of wind speed on CUHI magnitude, then the effects of other weather elements (e.g., cloud amount) must remain constant during the period of observation so as to highlight the role of wind. This can be achieved through experimental design or pre-processing of field data.

The interfering effects of topography—mainly surface relief, elevation, and water bodies—on T_a can be dealt with in a similar way. These effects are difficult to avoid in most cities, but some attempt to recognize and/or remove them from measurements of CUHI magnitude is necessary. For computations of CUHI magnitude, temperatures measured at sites of similar elevation, slope steepness, slope orientation, surface relief, and distance to water bodies can help to isolate the *urban* influence, while reducing the confounding effects of *nonurban* influences. Likewise, temperatures measured in gullies, ravines, basins, valleys, or on hillslopes or mountainsides—all of which allow cold air drainage and ponding at night—can confuse one's computation and interpretation of CUHI magnitude. Sampling designs that are configured parallel—not perpendicular—to elongated terrain features such as valleys, ridgelines, and coastlines can further reduce the effects of surface relief (e.g., mountain-valley wind systems) and water bodies (land-sea breezes) on computations of CUHI magnitude. In these designs, one must always consider diurnal changes in the direction of local winds and small-scale flow patterns relative to (i) the urban area, (ii) the measurement sites, and (iii) the source regions of cold (or warm) air. All such considerations for establishing control of nonurban topography are represented in Fig. 4.7.

Alternatively, corrective measures can be performed on the data after field collection. Two such techniques might improve isolation of urban effects in complex terrain (Goldreich, 1984). The first technique regresses temperature against height to determine a characteristic lapse rate for a particular study area. The observed temperatures can then be normalized to a standard height-level using the measured lapse rates. The second technique regresses temperature against distance inland from a large water body to determine a characteristic water-land temperature profile for the study area. The variable effects of a water body on urban and rural temperatures can be reduced by normalizing the observed temperatures to a standard distance from the shoreline. However, both of these techniques have serious drawbacks (namely the instability of regression equations) and should be used cautiously, if at all, to correct estimates of CUHI magnitude.

Lastly, if temperatures used to quantify CUHI magnitude are not measured synchronously, or adjusted so as to be synchronous, *urban-induced* heat islands may

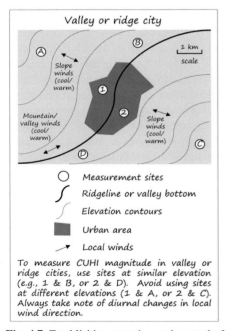

Valley or ridge city

- ○ Measurement sites
- ∫ Ridgeline or valley bottom
- ∫ Elevation contours
- ■ Urban area
- ↗ Local winds

To measure CUHI magnitude in valley or ridge cities, use sites at similar elevation (e.g., 1 & B, or 2 & D). Avoid using sites at different elevations (1 & A, or 2 & C). Always take note of diurnal changes in local wind direction.

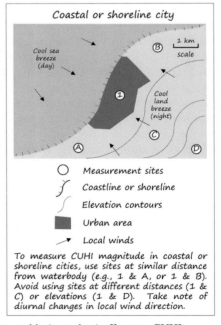

Coastal or shoreline city

- ○ Measurement sites
- ∫ Coastline or shoreline
- ∫ Elevation contours
- ■ Urban area
- ↗ Local winds

To measure CUHI magnitude in coastal or shoreline cities, use sites at similar distance from waterbody (e.g., 1 & A, or 1 & B). Avoid using sites at different distances (1 & C) or elevations (1 & D). Take note of diurnal changes in local wind direction.

Fig. 4.7 Establishing experimental control of topographic (nonurban) effects on CUHI magnitude.

be confounded with *time-induced* heat islands. This demonstrates the importance of time control, without which the resulting CUHI magnitudes might be erroneous. For mobile surveys lasting more than 1–2 h, temperature changes that occur during the survey period (due to a rise or fall in regional temperature) must be adjusted because those changes are not caused by urbanization. The method of adjustment is explained in Fig. 4.8. Temperature-time adjustments are normally made with simple bi-variate regression (temperature vs. time), with the slope of the trend line used to adjust the measured temperature values to a common base time. The rate of regional temperature change (cooling or heating) during a mobile survey can be calculated from the difference between the first and last measurement points of a closed traverse route (meaning that the traverse starts and finishes at the same point), or between one or more cross-over points along an open (or closed) traverse, with the points ideally located in different LCZ classes (Fig. 4.8). It is ordinarily assumed that the temperature change during a mobile survey is linear, but this is often not the case. Temperature records from fixed climate stations or automatic sensors within or near the study area can also be used for temperature-time adjustments. One must remember, however, that the temperature change—and thus the temperature adjustments—will differ between urban and rural environments, and even among local areas (i.e., LCZs) within those environments. Investigators must therefore consider the spatial representativeness of the sites used to adjust the temperatures. Rooftop sites are not recommended for temperature-time adjustments.

Time-control of stationary survey data is less problematic because the instruments can be programmed to record temperatures synchronously, in which case no temperature adjustments are needed. However, if temperature maxima or minima are used to quantify CUHI magnitude, one should be aware that these values are not normally synchronous across a spatial network of sensors, especially in urban and rural areas or in nonsteady weather. In this case, the maxima and minima should be adjusted to a common time based on temperatures recorded synchronously elsewhere in the study area, with either mobile or stationary surveys.

4.2.4 Quantifying the CUHI magnitude

To quantify CUHI magnitude, one needs to assign an operational definition to the term. An operational definition is one that translates concepts into field procedures. Arguably, the most reliable procedure to quantify CUHI magnitude involves urban–pre-urban temperature comparisons using only one urban measurement site. The site must not have moved since pre-urban times, despite changes to its surrounding landscape (Lowry, 1977). Unfortunately, such longitudinal temperature records are nearly impossible to find in most cities. The alternative, and more conventional, procedure is to measure synchronous T_a differences between urban and nearby rural (or nonurbanized) sites. This is expressed as ΔT_{u-r}. The operation, however, is problematic because the interpretation of its parts is variable and subjective: it may involve hourly, daily maxima/minima, or mean monthly or annual temperatures, and the classification of measurement sites as "urban" or "rural" is open to considerable bias. Nevertheless, studies of CUHI magnitude at short time scales might focus on certain hours of the day that correspond with routine human

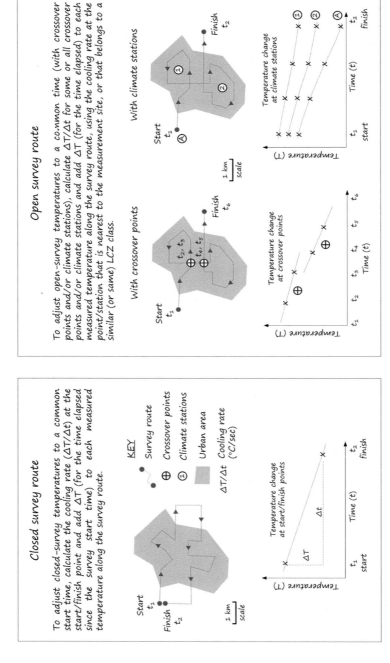

Fig. 4.8 Temperature-time adjustments for mobile CUHI surveys at night.

activities (e.g., sleeping, working, commuting), or at single hours that, in theory, are known to exhibit maximum (or minimum) thermal differences between urban and rural areas, or among LCZ classes.

The procedure to measure CUHI magnitude can therefore involve one of three possible comparisons: (i) a single urban site with a temperature record extending back to pre-urban times, allowing for a *temporal* comparison of urban and pre-urban temperatures; (i) a single pair of urban and nonurban sites, allowing for a *spatial* comparison of urban and nonurban temperatures; or (iii) several urban, nonurban, and/or pre-urban sites, allowing for comparison of spatially and/or temporally averaged temperatures. In this framework, a very common approach is to use multiple rural sites positioned on opposite sides of the city, and to combine the temperatures into a single mean value for comparison with one or more urban sites. Alternatively, one might select a rural site that consistently records low temperatures, but only if the site is representative of its local area, and if the elevation, surface relief, and distance to large water bodies is comparable to that of the urban area (see Fig. 4.7). If just one rural site is available, this may be the single option to represent the nonurbanized landscape. In all such cases, one must keep in mind the direction of prevailing winds in the region, and the shifting pattern of downwind urban effects on T_a. In Fig. 4.9, it is indicated that over the short term (hours to days), the urban-affected area extends downwind of the city due to horizontal transport of warm city air; when averaged spatially over longer periods (months to years), the area encompasses all wind directions and surrounds the city. If winds are calm, the urban-affected area will coincide with the urban area itself. The same principles apply to sites in the city: you might use one site in the city's central core, where nighttime temperatures are often highest, or a spatial average of representative sites at other positions in the city that are classified according to LCZs.

In the end, operational definitions will depend upon the time and space dimensions of the study: long-term comparisons (years or decades) of CUHI magnitude for urban-rural station pairs (at representative sites) will reveal the sensitivity of the heat island to changes in daily weather, seasonal soil moisture, and urban expansion, whereas a spatial average of many urban and rural sites might reveal intra-urban and intra-rural temperature effects on CUHI magnitude. Likewise, if the *maximum* magnitude is sought, it should be measured at a compact urban site (LCZs 1–3) in the core of the city, and at a bare or vegetated rural site (LCZs A–F) in the countryside and away from any urban effects extending downwind of the city (Fig. 4.9). Temperatures should then be recorded at night, several hours after sunset, and during calm, clear, and dry weather.

A variation to the ΔT_{u-r} approach, and one which provides a more purposeful definition of heat island magnitude, is to measure temperature differences between two or more sites classified according to LCZs (Stewart et al., 2014). This would require classification of rural or suburban sites into the "land cover" series (LCZs A–G), and city sites into the "built" series (LCZs 1–10). The definition is then expressed as a temperature difference between LCZ classes, e.g., $\Delta T_{LCZ\ 2-D}$, rather than as a difference between "urban" and "rural" classes, or ΔT_{u-r}. The advantage of this approach is that the physical and functional characteristics of the sites are explicit in their LCZ designations, which helps to communicate the essential metadata of an urban temperature study. It also allows for intra-urban temperature comparisons if several sites within the

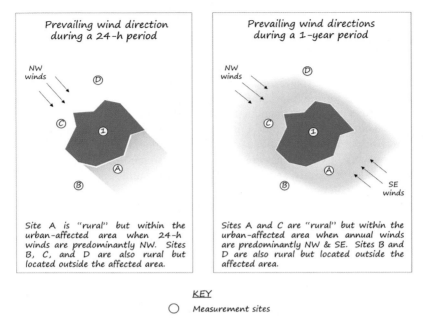

Fig. 4.9 Urban temperature effects extending downwind of a city.

city are assigned to LCZ classes. One can then make comparisons between, for example, highrise and lowrise zones, compact and open zones, commercial and recreational zones, and so on. This approach will not represent the heat island magnitude in a traditional sense because most (or perhaps all) of the sites will be positioned *within* the "island" itself. It will, however, reveal temperature variations at spatial scales smaller than the city (i.e., micro and local), which is both useful and necessary for many applications of heat island study (see, for example, Figs. 5.5 and 5.6). Finally, regular and repeated measurements over a period of at least several weeks to months is needed to establish control over random variation in all cases of heat island work, and to increase the probability of obtaining reliable estimates of the heat island magnitude and its internal temperature effects.

4.3   Concluding remarks

We have outlined the steps involved in planning an observational study of the CUHI. It is important to follow these steps in their proper sequence—this will ensure consistency, accuracy, and completeness of your work. The first steps require you to consider the purpose and audience of your study, while later steps introduce you to the methods, instruments, and definitions of heat island measurement. We conclude with a simple checklist to summarize the planning process.

1. *Ask questions*
   a. What is the problem, issue, or curiosity you wish to investigate?
   b. What resources are available to you (e.g., time, equipment, finances)?
   c. Who is your audience (e.g., scientists, policy makers, lay citizens)?
2. *Choose a study method*
   a. Select mobile and/or stationary surveys. Be sure to align these with the aims and resources of your study.
   b. Determine the instruments needed, and keep in mind their technical specifications.
   c. Follow international guidelines to install and house the thermometers.
3. *Design a sampling framework*
   a. Identify survey sites and routes that are representative of the local scale. LCZ maps and site visitations are essential in this process.
   b. Determine sampling frequencies and density of measurements. These must align with the time and space scales of your study.
   c. Configure the survey sites to control for unwanted effects on CUHI magnitude (e.g., weather, surface relief, water bodies).
4. *Define the CUHI magnitude*
   a. Choose definitions that are relevant to your audience. Consider city scale (ΔT_{u-r}) and local scale (ΔT_{LCZ}) indicators.
   b. Evaluate the need for spatial and/or temporal averaging in your definition.

References

Aguilar, E., Auer, I., Brunet, M., Peterson, T.C., Wieringa, J., 2003. Guidance on Metadata and Homogenization. WMO Technical Document No. 1186, World Meteorological Organization, Geneva.

Goldreich, Y., 1984. Urban topoclimatology. Prog. Phys. Geogr. 8, 336–364.

Lowry, W.P., 1977. Empirical estimation of the urban effects on climate: a problem analysis. J. Appl. Meteorol. 16, 129–135.

Meier, F., Fenner, D., Grassmann, T., Otto, M., Scherer, D., 2017. Crowdsourcing air temperature from citizen weather stations for urban climate research. Urban Clim. 19, 170–191.

Oke, T.R., 2006. Initial Guidance to Obtain Representative Meteorological Observations at Urban Sites. IOM Report 81. World Meteorological Organization, Geneva.

Stewart, I.D., 2011. A systematic review and scientific critique of methodology in modern urban heat island literature. Int. J. Climatol. 31, 200–217.

Stewart, I.D., 2019. Why should urban heat island researchers study history? Urban Clim. 30, 100484.

Stewart, I.D., Oke, T.R., Krayenhoff, S.E., 2014. Evaluation of the 'local climate zone' scheme using temperature observations and model simulations. Int. J. Climatol. 34, 1062–1080.

Sundborg, Å., 1951. Climatological studies in Uppsala with special regard to the temperature conditions in the urban area. In: Geographica. 22. Geographical Institute of Uppsala, Sweden.

Szymanowski, M., 2005. Interactions between thermal advection in frontal zones and the urban heat island of Wrocław, Poland. Theor. Appl. Climatol. 82, 207–224.

WMO (Ed.), 2008. Guide to Meteorological Instruments and Methods of Observation, seventh ed. World Meteorological Organization, Geneva. WMO-No. 8.

Executing a CUHI study

5

After the fieldwork of a CUHI study has been completed, a large dataset of air temperatures will have been amassed. The task now becomes the systematic organization, management, and documentation of those data for their intended audience. In this chapter, we guide you through this task and identify five key steps: (1) compiling metadata, (2) quality control of metadata, (3) stratifying metadata, (4) processing metadata with heat island observations, and (5) illustrating the heat island observations. The purpose of these steps is to ensure, above all else, that the measured temperatures in your city have been carefully scrutinized for use in a scientific study of the urban climate, and that calculations of the heat island effect are reliable and robust. The extent to which you follow these steps will depend upon the aims and audiences of your work. If it is targeted at landscape designers, the metadata you collect—and the criteria by which you stratify them—will include local and microscale descriptors of surface geometry and building materials, rather than regional cloud or soil moisture (the latter being of more interest to meteorologists, for example). The first of the five steps—compiling metadata—is paramount to this chapter and to all types and purposes of heat island investigation. Fundamentally, the compilation of metadata prompts you to think critically about the methodological and technological constraints of your instruments, and to analyze and interpret the observed heat island effect within those constraints. This will lead you to accurate conclusions and valid comparisons of your work to that of others.

5.1 Compiling metadata

The term *metadata* is defined as "information about the data." In a CUHI study, metadata comprise the qualitative and quantitative information about the air temperature (T_a) readings: how, when, where, and by whom the readings were taken, and under what environmental and experimental circumstances. The WMO requests that this information be reported for all meteorological observations, to include the station history, its geographic location, instrument specifications, and local environment, as well as the observing practices, recording procedures, and data-processing methods (Aguilar et al., 2003). Supplementary guidelines for meteorological observations in *urban* areas stress that metadata on the history and local environment of the stations (or field sites) are essential to CUHI studies due to the complex and changeable landscapes of the urban environment (Oke, 2006a,b). Also important are metadata on the regional weather conditions during a heat island survey. While all such data are important to the analysis and interpretation of the CUHI, it is unlikely that these data can (or will) be sourced and documented in full, particularly for studies involving

The Urban Heat Island. https://doi.org/10.1016/B978-0-12-815017-7.00005-9

networks of dozens, or hundreds, of observing stations and measurement sites. In fact, full documentation would complicate a metadata file to the extent that operational uses of the information would be restricted; in other cases, metadata on station history or synoptic weather may not exist or may not be accessible to the investigator. Therefore, the metadata that we describe and recommend here are those deemed to be most accessible to a basic investigation of the CUHI, and most essential to its quality. We sort these metadata into four groups: (1) instrument type, mounting, and shielding; (2) observing practices and schedules; (3) local environment of the measurement site; and (4) regional weather during the observation period.

Our emphasis on metadata in this chapter is rooted in the regrettable fact that very few heat island investigators communicate this information in their published reports (Stewart, 2011). This has meant that the measured quality of the heat island estimates in the literature is generally poor or unknown, and thus the usefulness of the estimates is diminished. We therefore urge you to take the collection of metadata seriously: it should not be seen as a tedious undertaking, but instead as an intriguing search for information that helps you to produce a more complete assessment of the heat island effect. Maps, photographs, tables, charts, graphs, logbooks, field notes, and field sketches are all helpful sources and mediums through which to collect and convey basic metadata.

5.1.1   Instruments

Metadata on the instrumentation of a CUHI study must specify the type of instruments used, and the mounting and shielding of those instruments during the observation period. "Instrument type" describes the category of thermometer that is used (e.g., electrical, liquid-in-glass), to include the model and manufacturer. Metadata must also specify the instrument's ventilation (natural or forced), its measurement resolution (i.e., the smallest change in temperature that can be detected by the thermometer), and its uncertainty (the interval of values within which the true temperature lies). Accompanying these metadata are the instrument's sampling intervals (e.g., minutes, hours), response times (or time-constants), and time averaging (if applicable).

Instrument "mounting and shielding" refers to the height placement of the thermometer above ground level; the size and style of shield or screen enclosing the thermometer (see Fig. 4.2); and the degree to which nearby surface facets, elements, or terrain features interfere with the measured T_a values. Whether or not an interference is deemed "erroneous" will depend upon the desired scale and representativeness of the temperature measurements. Possible interferences include an irrigated lawn or garden; a fence or instrument mast; an individual tree or automobile; a surface dip, hollow, or hill in the terrain; or a building window, doorway, or HVAC unit. Also needed is a detailed description (including dimensions) of the supporting surface or structure to which the instruments (and their shields) are attached, and whether that surface or structure is mobile or stationary. Common examples of stationary structures include lamp posts, traffic signals, building walls, and building rooftops. Mobile structures include bicycle frames and vehicle doors, roofs, or bumpers (see

Fig. 5.1 Photographic metadata for a Stevenson screen in rural Hong Kong.
From I.D. Stewart.

Fig. 4.4). If the instrumentation is changed during the observation period of a CUHI study, the date of change must be noted in the metadata files, along with specifications for both the original and replacement instruments, including their mounting and shielding. Much of the these data can be represented in a heat island report with a set of photographs illustrating the thermometer shield and its position at the measurement site, along with a table containing information on the siting conditions of the instruments. Photographs should be taken from several points of the compass, and at sufficient distance to expose the mounting apparatus and its background setting (Fig. 5.1).

5.1.2 Observing practices

Metadata on observing practices and schedules is particularly important for long-term CUHI studies, which usually involve stationary rather than mobile surveys. Important to all studies, however, are the times, dates, and frequencies at which the temperatures are measured—for example, every second, minute, hour, or day, or at predetermined times that coincide with temperatures of special interest (e.g., daily maxima or minima). The start and finish times (and thus duration) of a field campaign and its temperature surveys must also be included in the metadata files, along with times and dates of interruption or change to the observation schedules and practices. Such interruptions will inevitably occur in studies lasting for many months or years, and will break the continuity of the temperature-time series and reduce its quality, both of which require acknowledgment and/or correction. Any such corrections applied to the temperature data must be documented in the metadata files, along with site inspections and routine field checks for instrument comparisons and maintenance (e.g., cleaning, calibration, fault detection).

5.1.3   Measurement sites

Air temperatures at standard screen height (1.25–2 m above ground) are influenced by many factors that act at overlapping spatial scales, namely micro, local, and regional. Each of these scales is relevant to the metadata requirements for "local environment," which describes (quantitatively and qualitatively) the surface structure, cover, fabric, and metabolism of the site and its surrounding turbulent source area (several 10s to 100s of meters in the horizontal) (Oke, 2006a,b). *Structure* refers to the dimensions (e.g., height and spacing) of the main surface elements in the source area, such as trees and buildings; *cover* refers to the surface type (paved, vegetated, soil, water) in the source area; *fabric* refers to the material properties of the cover (albedo, heat capacity, thermal conductivity); and *metabolism* refers to the human-generated emissions of heat, water vapor, and air pollutants. The simplest and most encompassing way to document these characteristics of the local environment is to categorize the sites according to Local Climate Zones (LCZs) (Section 2.3.2; Tables 2.3 and 2.4). Each LCZ class is associated with a set of parameters that describe the local environment for its physical structure (e.g., sky view factor, aspect ratio, building and tree heights), land cover (surface fractions), surface fabric (thermal admittance—combining the properties of heat capacity and thermal conductivity), and urban metabolism (anthropogenic heat flux). Categorizing a site into its best-fit LCZ class should involve estimates of these parameters. This is a worthwhile pursuit because it allows a heat island investigator to establish the essential metadata for some or all of the sites used in the study. If *measured* values are not necessary or not available to the study, one should at least estimate *qualitatively* the most basic LCZ parameters, using visual observations of the field site, whether in situ (via personal visits to the site) or remotely (via maps, photographs, or satellite images). All such estimates will be open to personal judgment, but even this is preferable to a complete lack of environmental metadata. Fig. 5.2 presents a general template to be used for documenting local environment metadata for a CUHI measurement site. The template features a local sketch map (which can be digitized, hand-drawn, or photographed), alongside essential metadata for the local and microscale setting of the site.

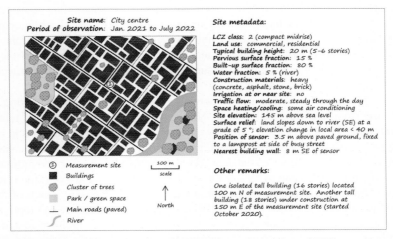

Fig. 5.2 Template for documenting local environment metadata for a CUHI measurement site.

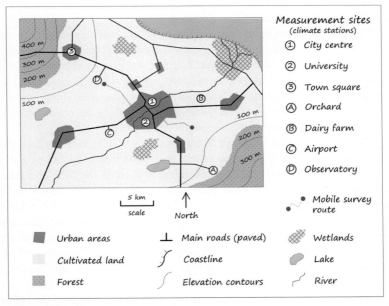

Fig. 5.3 Regional map for a CUHI study.

Not to be excluded from the metadata on "local environment" is a map or aerial photograph situating the study area in its regional context. As shown in Fig. 5.3, the map must portray the major divisions of the region, such as mountains, coastlines, hills, valleys, basins, and water bodies, as well as urbanized areas, transportation routes, agricultural land uses, natural vegetation, and other terrain features relevant to the climate. Also to be indicated in the map are the sites of temperature measurement, whether these belong to a mobile survey or a network of sensors. Accompanying this (or another) map should be a metadata table containing the locations, elevations, slope characteristics (steepness, orientation), and exposures (sky view factor) of the measurement sites, as well as the ground or surface cover directly beneath the instruments (e.g., asphalt, grass, bare soil).

For CUHI studies involving a small network of observing stations, gathering metadata on the siting environment of each station is certainly manageable and relatively straightforward. In this case, we strongly encourage you to visit the stations in person to observe firsthand the position of the temperature sensor, the layout of the site, and the natural and built features of the area. You can then make an informed judgment as to the suitability of the station for your heat island study. If you are working remotely with the stations, or if access to the stations is restricted or unsafe, much of the environmental metadata can be extracted from secondary sources such as topographic maps, satellite images, aerial or ground-level photographs, or archival materials in local libraries or weather offices. Geobrowsers such as *Google Earth* and *Google Maps* are helpful for retrieving metadata remotely. Ideally, however, these secondary sources should not be used as a substitute for primary sources from personal visits to the stations. The value of such visits for sourcing local environment metadata (even if qualitative) cannot be overstated.

For CUHI studies involving dozens, or hundreds, of observing stations or measurement sites, gathering a complete set of local environment metadata is a daunting, and perhaps an impracticable, task. However, full metadata on local environment for all such sites is arguably unnecessary because many of the sites will co-exist in a shared neighborhood and/or LCZ class. Metadata for some of these sites should therefore not differ greatly, in which case it may be sufficient to document the information for only a reduced number of representative sites in the study area.

Naturally, if the location of one or more sites in a CUHI study is changed during the observation period, its local environment will also change. Metadata files should therefore be created for both old and new sites, to include the time and date of the move, the locations of the sites, and their surface and exposure characteristics. This information is essential for historical heat island studies lasting many years—or of heat island datasets covering many decades—during which time site relocations occur and operations are interrupted. Similarly, if the local surroundings of a climate station change dramatically during the period of a CUHI study, owing to the growth, renewal, decline, or destruction of a neighborhood or plot of land, this too needs to be documented. Over time, most urban environments experience some degree of change caused by road and building construction (or demolition); bedding (or removal) of trees and plants; and/or introduction of new (or termination of old) transportation routes, irrigation schedules, and so on. These changes to the physical environment of a measurement site will influence the observed temperatures at that site. Its metadata files must therefore be updated regularly. Similar changes can occur at rural sites if forests or fields are cleared, replaced, or rotated with new land covers. At shorter time scales, seasonal changes to the local environment of urban and rural sites will coincide with rainfall and snowfall distribution patterns, crop seeding and harvesting schedules, and the timing of bud burst and leaf fall for deciduous trees.

5.1.4   Regional weather

Metadata on regional weather during the observation period are required of any empirical CUHI study because air mass properties exert a controlling influence on surface climates. Meaningful analysis and interpretation of the CUHI therefore requires full and accurate weather data. At minimum, investigators should provide metadata for wind (speed, direction), cloud cover (type, amount, base height), and precipitation (type, amount). These data are needed to characterize the air mass during the observation period, and to deduce the moisture status of soils in the region, which strongly influences surface heating/cooling rates and near-surface air temperatures. For some applications, additional weather data such as humidity and atmospheric pressure might be useful; however, these weather elements are known to exert less influence on CUHI formation.

The representativeness of weather metadata must be regional in scale. This is because CUHI magnitudes are highly sensitive to synoptic winds and cloud cover. Regional data can be sourced from official weather stations, often located near airports or on the outskirts of cities. The purpose of these stations is to observe air mass properties for synoptic weather forecasts, not to monitor heat island effects for local or microscale climate studies. The observations therefore must comply with WMO guidelines for synoptic stations, such that measurements of wind speed and direction represent the climate region to 10s or 100s of kilometers around the station. For this to happen, wind sensors must be placed 10 m above ground level (not at screen height or on building rooftops) in open, relatively flat terrain with surface obstructions such as trees, buildings, and fencing no closer to the sensor than ten times their height. If these data cannot be obtained from a synoptic station, noninstrumented observations of wind, cloud, and soil moisture during a survey can be a source of replacement data. For example, wind speed and direction can be estimated with the standardized Beaufort scale (Table 5.1), which requires your own sensory observations of wind—what you can see and feel—at an open location away from eddies and gusts caused by buildings. The Beaufort scale was created in 1806 for use at sea, but is now widely employed for estimating wind and its effects on land objects, e.g., smoke drift from tall chimneys, movement of leaves and tree branches. However, the movement of clouds overhead must not be used for estimating wind at ground level. Other sensory observations based on sounds and smells might also be helpful in the absence of instruments. For example, traffic sounds are often carried by wind from nearby sources such as airports and railroads, and environmental odors drift downwind from factories and farms. Paying careful attention to your sensory experiences will give you clues about the speed and direction of local and regional winds.

In the absence of synoptic data, estimates of *cloud amount* can be made by visual observations of the "sky dome" overhead. These observations are difficult to perform at night, but made somewhat easier if the site is not overly polluted with fog, dust, smoke, haze, or urban lighting, or otherwise surrounded by mountains or landforms that obscure the horizon. While at the measurement site, tilt your head back and from side to side to observe the sky in its entirety, covering the complete horizon. If the sky is sufficiently dark, your estimate of cloud amount will be based on the proportion of stars that are dimmed or completely hidden by clouds. This should enable a characterization of cloud cover as "partial," "total," or "no cloud"; if conditions

Table 5.1 Beaufort scale for estimating wind speed over land. Scale numbers higher than 9 are not listed.

Beaufort scale number	Description	Wind speed equivalent at a standard height of 10 m above open, level terrain		Specification for estimating wind speed over land
		m s⁻¹	km h⁻¹	
0	Calm	0–0.2	<1	Calm; smoke rises vertically
1	Light air	0.3–1.5	1–5	Smoke rises nearly vertically; direction of wind shown by smoke-drift but not by wind vanes
2	Light breeze	1.6–3.3	6–11	Wind felt on face; leaves rustle; ordinary vanes moved by wind
3	Gentle breeze	3.4–5.4	12–19	Leaves and small twigs in constant motion; wind extends light flag; hair is disturbed; clothing flaps
4	Moderate breeze	5.5–7.9	20–28	Wind raises dust and loose paper; small branches are moved; hair disarranged
5	Fresh breeze	8.0–10.7	29–38	Small trees begin to sway, crested wavelets form on inland waters; force of wind felt on body
6	Strong breeze	10.8–13.8	39–49	Large tree branches in motion; whistling heard in power lines; umbrellas used with difficulty; some inconvenience in walking
7	Near gale	13.9–17.1	50–61	Whole trees in motion; difficulty when walking against wind
8	Gale	17.2–20.7	62–74	Twigs break off trees; difficulty with balance when walking; wind generally impedes progress
9	Strong gale	20.8–24.4	75–88	Slight structural damage to building roofs (e.g., tiling blown off); people in danger of being blown over

Source: WMO, 2008. Guide to Meteorological Instruments and Methods of Observation. seventh ed. WMO-No. 8. World Meteorological Organization, Geneva.

permit, you might attempt a more quantitative assessment of the sky fraction covered by cloud (expressed in tenths), with 0 and 1 signifying completely clear and overcast skies, respectively. You might also attempt to identify the *cloud type* as cirrus, cumulus, or stratus, and the cloud-base height (above ground) as high, middle, or low level. All of these observations can be helped by cloud identification guides that are available from many sources. The WMO (2017) *International Cloud Atlas*, for example, provides step-by-step instructions for making internationally comparable cloud observations of the amount, type, and base height of cloud cover (daytime or nighttime) for observers located at ground level and without measuring equipment. A summary of cloud information for analyzing CUHI data is provided in Table 5.2. Finally, ground moisture at the measurement sites and in the surrounding countryside can be described with simple terms such as "dry," "moist," "wet," or "flooded," based on the look and

Table 5.2 Information on common cloud types.

Cloud type[a] (genus)	General appearance	Height category	Height of cloud base above ground level	Cloud coefficient[b] (k)
Cirrus	Clouds detached in narrow bands; silky, hair-like appearance	High	5–18 km	0.16
Cirrostratus	Clouds thin with whitish veil across sky; smooth appearance	High	5–18 km	0.32
Altocumulus	Small clouds with rounded heaps; patchy or sheet-like appearance	Middle	2–8 km	0.66
Altostratus	Sheet or layered cloud across whole sky, thin in parts; uniform appearance	Middle	2–8 km	0.80
Cumulus	Small, detached clouds with dense, puffy heaps and sharp edges	Low	0–2 km	0.80
Stratocumulus	Patchy sheet or layered cloud with rounded heaps, merged or slightly open	Low	0–2 km	0.88
Stratus	Dense, layered cloud with uniform base; featureless appearance	Low	0–2 km	0.96
Fog	Stratus-like cloud with base at ground level; visibility less than 1 km	Surface	0	1.00

[a] Luke Howard, the pioneer of urban heat island studies (Chapter 1), invented the first cloud-naming system in the early 19th century.
[b] Used for calculating the weather factor (ΦW) in Section 5.3.3.
Sources: Oke, T.R., 1987. Boundary Layer Climates. Routledge, London. WMO, 2017. International Cloud Atlas. Manual on the Observation of Clouds and Other Meteors, vol. 3. WMO-No. 407. World Meteorological Organization, Geneva. Available at https://cloudatlas.wmo.int/home.html.

feel of the soils (e.g., cracks in the ground; dust or loose sand on the surface; standing water in small or large pools). Admittedly, this approach is not always accurate (due to observer bias), but it is easily carried out in any field study.

The desired time-resolution of regional weather data will depend on the type of CUHI study being conducted. If the study involves mobile surveys, the metadata should be obtained for the observation period of each survey (typically 1–2 h in duration), as well as for the hours preceding the survey. This is necessary because wind, cloud, and rainfall can suppress the formation of heat islands and local climates, well before their magnitudes are recorded in the field. For a nocturnal survey beginning 3–5 h after sunset (i.e., the usual time of daily maximum CUHI magnitude), "preceding" refers to the hours between sunset and the start of the survey. Rainfall metadata may be required for several days preceding a temperature survey, especially if the rain is prolonged or intense and/or the soils and surfaces are poorly drained and are wet during the survey. Cloud, wind, and precipitation data at hourly intervals are ideal for analysis of mobile temperature surveys, but if these data are not available at that frequency, then three-hourly or daily intervals can be helpful. For long-term CUHI studies, hourly weather data are preferred, but longer time-intervals will suffice for most purposes.

5.2   Quality control of metadata

As a small measure of quality control, it is advantageous to examine the temperature and metadata files that you have amassed. This allows you to identify and remove inaccurate, unwanted, uncertain, or implausible measurements from further analysis, and to gain confidence and consistency in your results. Quality control may identify flawed or erroneous measurements associated with (i) malfunctioning instruments (which lead to "instrumental errors"); (ii) human oversight or mistaken recordings ("gross errors"); and (iii) peculiar changes to the external environment of the measurements ("environmental errors"). All such errors are inevitable but their overall impact can be kept small if the field operations are done carefully and the instruments are exposed properly. This sentiment is clearly stated by the WMO (2008): *Measured data are imperfect, but, if their quality is known and demonstrable, they can be used appropriately.*

A first step toward quality assurance (and data reduction) is to remove from analysis—or flag as unreliable—data obtained from sites that are not suitable for CUHI studies. These are sites that are poorly located according to WMO guidelines for urban or rural meteorological observations (Oke, 2006a; WMO, 2008). Data gathered from a building rooftop, for example, may be rejected if the temperatures are judged to be unrepresentative of the environment below roof level where the operations of the study are meant to be carried out. Likewise, data from instruments fixed too closely to building walls, windows, or doorways may be removed (or flagged) if their exposures are thought to generate unacceptable radiation or ventilation errors. These errors are especially common among volunteer and crowd-based networks of temperature sensors. The researcher, therefore, needs to exercise a good measure of quality control when using crowd data in an urban temperature study. This can be done by removing temperatures that lack essential metadata, or otherwise appear erroneous due to calibration problems, software errors, instrument design flaws, or nonrepresentative siting. The remaining data should then be crosschecked against high-quality temperature data from official stations nearby. Alternatively, one may wish to visit and inspect the measurement sites before rejecting or accepting their use in a CUHI study.

As a general rule, we recommend removing all data not representative of the desired length scale (e.g., local or city), or not meeting the stated or implied goals of the work (e.g., to measure ΔT_{u-r} or $\Delta T_{LCZ\,x-y}$). Temperatures influenced by extraneous microclimatic effects should also be removed: these include heat plumes, frost pockets, cold air run-off, land-water boundaries, and uncharacteristic surfaces and objects near the instruments. Exceptions will arise, however, if the effects are deemed relevant to the novelty or goals of the study, or if the temperatures influenced by these effects are acknowledged for their potential errors (with supporting metadata). For an automobile survey, temperatures obtained while the car stops or slows at a street intersection—thereby exposing the thermometer to sources of heat that may not be typical of the local area—can be described as "poor quality" and flagged for removal from the study. Similarly, temperatures observed from bridges, tunnels, overpasses, and underpasses, or next to isolated trees or buildings, during the survey should be carefully scrutinized for consistency with the scales and aims of the study. If the scales of observation are not compatible with the higher aims of the study, the associated data should either be eliminated or acknowledged for their inconsistencies. If, however, the data help to meet a specific aim of the study, or otherwise to demonstrate a constructive lesson, argument, or result, their inclusion can be justified.

5.3 Stratifying metadata

The steps by which metadata are stratified for a CUHI study follow those of Section 5.1—in other words, the metadata are sorted into smaller files according to time (e.g., hour, day, season), site characteristics (surface structure, land cover, LCZ class), and weather conditions (wet/dry, calm/windy, warm/cold). The purpose of stratification by these criteria is to isolate the causal or influential factors behind the observed CUHI magnitudes and their changes in time and space. Stratification thus yields an element of "control" (passive, not active) over these factors. This allows you to identify, set apart, and later examine the factors of most importance to your work. If, for example, your work is to investigate the efficacy of green or blue infrastructure (e.g., parks or small lakes) to reduce urban heat at the neighborhood scale, then stratifying the site metadata according to LCZ class and surface cover fractions can help to establish meaningful relations between air temperature, tree density, and land cover. Alternatively, if your work is to inform utility managers or urban gardeners of the meteorological influences on CUHI magnitude, then stratifying the data according to sky and soil conditions will lead to more interesting and worthwhile communications.

5.3.1 Time

The appropriate time scale for stratification will be determined partly by the duration of the CUHI study. For studies lasting several days or weeks, stratification will occur at hourly and daily intervals, so as to separate the metadata into time strata that correspond to daytime/nighttime periods and individual hours of the day (normalized to the time of sunrise/sunset). This is helpful because the magnitudes of heat islands and local temperature differences will change markedly within this range of scales. Do not feel discouraged if you have only a few temperature observations for each hour; instead, understand that the main intent of local and microclimatic research is to investigate specific times and weather conditions that generate sharp thermal gradients across vertical and horizontal distances of only a few 10s or 100s of meters. For lengthened time scales of several seasons or years, stratification will occur at hourly, daily, monthly, and/or yearly intervals. The strata then reveal temperature signals that fluctuate with time of day (relative to surface heating and cooling cycles), time of week (relative to human activity cycles), and season of the year (relative to solar angles and seasonal precipitation patterns). For longer time scales of years to decades, urban-rural stratification is necessary to separate CUHI effects from those of global warming or background climate change in the temperature record (Lowry, 1977). The assumption in this approach is that the rural temperature signal has been affected only by global warming, while the urban signal has been affected by both urban *and* global warming. Subtracting rural temperatures from nearby urban temperatures should give a reasonable estimate of the CUHI effect without contamination by background climate change, so long as the T_a values have been measured synchronously and in the same climatic province.

5.3.2 Site characteristics

To isolate the effects of land cover, soil moisture, urban structure, or other physical characteristics of the measurement environment, it is necessary to stratify the metadata according to narrower categories of special interest, and to seek corresponding changes in T_a and CUHI magnitude. The strata might include one or more of the standard class properties of the LCZ scheme, viz.: building height (e.g., highrise, midrise, lowrise), building spacing (compact, open), and surface cover (pervious, impervious). Values to define these properties are provided with the LCZ scheme, which gives a characteristic range of values that is applicable to all cities (Table 2.4). You may wish to set new ranges or values that are more suitable to your study area or study objectives.

Site stratification is important because it affords a degree of experimental control over the variables of concern to your study. If building height or canyon aspect ratio is of most interest, look for temperature changes among the measurement sites and their corresponding strata (e.g., highrise, midrise, lowrise), while keeping other critical characteristics of those same sites relatively constant. Any observed changes in T_a can then be attributed to the effects of varying building height or aspect ratio. Keep in mind that the effects of other confounding factors will always be present in the data because "active" control over all external influences is not possible in a field study. The purpose of stratification is only to *reduce* the effects of these external influences and subsequently to isolate cause-effect relations. In this respect, studies with a greater number of field sites (e.g., mobile surveys) have an advantage over those with a smaller number (stationary surveys) (Table 4.1), mainly because a larger sample size (i.e., number of sites) provides a more robust statistical relation between urban form and urban air temperature.

5.3.3 *Weather conditions*

To identify the maximum strength of a CUHI, or to spatially differentiate its local temperature zones, it is *essential* to stratify the temperature data by weather conditions; this is partly met by time stratification (Section 5.3.1). Weather stratification helps to isolate the sky (and soil) conditions that force heat islands and local climates to reach their clearest expression in the temperature record. These conditions include calm or light winds, clear or partly cloudy skies, and dry soils. Isolating such conditions will uncover interesting causal relations in the data, allowing attribution of the observed CUHI to *urban* effects, with little interference from *nonurban* factors such as precipitation, strong advection, or passing weather fronts.

The most effective way to stratify weather metadata for a CUHI study is to use the weather factor (Φ_W), as described by Runnalls and Oke (2000). Φ_W is derived from nonlinear relations between UHI magnitude and cloud and wind conditions:

$$\Phi_W = u^{-1/2}\left(1 - kn^2\right) \tag{5.1}$$

where $u^{-1/2}$ is an empirical relation between heat island magnitude and regional wind speed ($m\,s^{-1}$), and $1 - kn^2$ is the Bolz correction for net longwave radiation under a given cloud type (k, see Table 5.2) and cloud amount (n, in tenths on a scale from 0 to 1). While many researchers do consider cloud amount in their analyses, very few consider cloud type, which is a critical measure of the influence of cloud on near surface air temperatures. A large amount of high-altitude cloud (e.g., cirrus) is relatively unimportant to CUHI or local climate formation because its base temperature is low; on the other hand, a small amount of low-altitude cloud (e.g., stratus or cumulus) is powerful because its base temperature is high (Table 5.2).

Essentially, Φ_W expresses the extent to which wind and sky conditions permit nocturnal cooling of the surface layer atmosphere. It is treated as a nondimensional number able to indicate the relative cooling potential of the surface, with values ranging from 0 (poor cooling, no CUHI effect) to 1 (excellent cooling, maximum CUHI effect). Due to its square root form, the relation is limited to wind speeds $> 1\,m\,s^{-1}$, which is the default value in calm or near-calm conditions. Φ_W values can be grouped into desired ranges or intervals, which are used as a basis to stratify the temperature data of a CUHI study, with the highest values associated with the greatest CUHI magnitudes, assuming little or no precipitation. This procedure requires weather data for the time of the survey and/or the hours preceding the survey.

To isolate the singular effects of cloud, wind, atmospheric pressure, or other meteorological variables on CUHI magnitude, further weather stratification is needed. If wind speed is the variable of interest for its effect on CUHI magnitude, then the temperature data must be sorted into wind speed categories, with each category corresponding to a same or similar atmospheric state (cloud, humidity, etc.). Any changes in CUHI magnitude, whether expressed as $\Delta T_{u\text{-}r}$ or $\Delta T_{LCZ\,x\text{-}y}$, can then be attributed to the effects of changing wind speed. The same approach is used for other variables of interest such as cloud amount or cloud type. However, if the data are not stratified appropriately, it becomes difficult to extract valid relations from among the many meteorological influences on CUHI magnitude. To seek and study these relations, field surveys of longer duration are preferred over those of shorter duration, simply because the sample size (number of repeated observations) is larger and the statistical results will be more stable. Networks of fixed stations are immediately favored over mobile surveys for studying these relations (see Table 4.1).

5.4   Processing the data

After the metadata of a CUHI study have been compiled, quality controlled, and stratified into diagnostic groups, the next step is to calculate urban-rural and intra-urban temperature differences. The methods to do this have been discussed in Chapter 4, and generally involve either or both of two computations—$\Delta T_{LCZ\,x\text{-}y}$ or $\Delta T_{u\text{-}r}$—depending on the scale and objectives of the study. For most heat island studies, there is a basic need to establish statistical relations between the temperature data and measures of regional weather and/or urban form and function. Two common statistical approaches to meet this goal are correlation analysis and simple (or multiple) regression analysis. We will discuss both approaches and their use in CUHI studies. We will also discuss the use of site metadata to run a simple numerical energy balance model. Lastly, numerical temperature indicators can be used to express the thermal conditions of cities and their LCZ classes. These are easily relatable to the daily lives and occupations of urban inhabitants, especially to address issues of public health, settlement planning, and urban energy demand.

5.4.1 Correlation analysis

Correlation analysis measures the association between two or more variables by examining the covariation in recorded values, that is, the extent to which variation in one variable corresponds with variation in another. Although the technique does not distinguish among the variables under scrutiny, in UHI studies the aim is to determine the statistical relation between magnitude (as measured by $\Delta T_{u\text{-}r}$ or $\Delta T_{LCZ\,x\text{-}y}$) and variables relating to urban form, urban function, or regional sky conditions. In this framework, we identify $\Delta T_{u\text{-}r}$ as the dependent variable that responds to changes in the independent variables used to describe the urban landscape and weather. The strength and direction (positive or negative) of this association is expressed using the coefficients of correlation (r) and determination (r^2), which together convey a statistical relation (not necessarily a causation) that exists between the variables. A "perfect" correlation of 1 indicates that all of the variation in ΔT_{u-r} is linked completely to variations in the other variable(s); similarly, a value of 0 indicates no association. The *statistical* significance of the results indicates simply whether any associations uncovered could have occurred by chance; this depends on the sample size and can be assessed if the assumptions that underpin the correlation technique are met. The "practical" significance of any statistical relations uncovered should be placed in a meaningful context, which is best done within the energetic framework discussed in Chapter 2.

It is important to select variables that are relevant to the physical or energetic processes of energy and water exchange in cities, and that generate the heat island effect and its local climates. City population, as a statistic, is probably the most popular (independent) variable in the urban climate literature, but its relevance to the surface energy balance is limited to the metabolic heat flux (Q_M) from human bodies, which is negligible at local/urban scales in all but the most densely populated cities. Frankly, population is merely a surrogate for more structural measures of urban form, such as sky view factor, impervious surface fraction, and canyon aspect ratio. These measures change with increases in city population over time: as tall and/or close-set buildings are erected and vegetated surfaces are sealed, the sky view factor decreases and impervious fraction increases. Thus, using correlation analysis to link ΔT_{u-r} or $\Delta T_{LCZ\,x-y}$ with local or microscale measures of surface form and land function is a more satisfying approach. Correlation analysis should also include variables related to surface relief, such as elevation and slope steepness or orientation. A direct benefit of this approach is that it makes the collection of metadata on local environment an unavoidable step in any CUHI investigation (Section 5.1.3).

Equally important to the urban energy and water balances of heat islands and local climates are the regional sky (and soil) conditions prevailing before and during a measurement period. For this reason, correlations between meteorological elements and CUHI magnitudes have been reported in the climate literature for many decades, and for hundreds of cities. The statistical and graphical form of the relation between wind speed, cloud cover (type and amount), and ΔT_{u-r} are universally known and transferable from place to place, such that cloud and wind both demonstrate strongly negative correlations with ΔT_{u-r}. Also included in these relations to describe and explain ΔT_{u-r} are measures of atmospheric humidity, atmospheric pressure, and atmospheric stability (vertical lapse rate), all of which are integral to the work of urban climate modelers, weather forecasters, and global climate scientists.

5.4.2 Regression analysis

A natural complement to correlation is regression analysis, which fits a mathematical function that explicitly measures the relation between a dependent variable to one or more independent variables. *Simple* linear regression determines the equation of a straight trend-line (with an intercept and slope) that represents the "best fit" of a dependent variable to one independent variable; *multiple* linear regression determines the relation between several independent variables and a single dependent variable. If the relations between variables cannot be captured by linear assumptions, polynomial regression techniques are used. The result of the regression analysis is an equation that expresses CUHI magnitude as a function of independent variables. By implication, one's use of the equation states that the CUHI magnitude is caused by processes directly associated with the independent variables. Regression equations are used to predict values for ΔT_{u-r} based on the value of the independent variable(s), or to evaluate the impacts of modifying certain aspects of the urban environment on the CUHI. In both cases, the predictions often extend outside the range of observations. Like correlation, regression analysis is based on a set of assumptions that, if met, allow the researcher to establish the statistical significance of the results and to express confidence-bounds around any predicted ΔT_{u-r} values.

In CUHI studies, use of regression analysis dates back to Sundborg's (1951) study of the heat island in Uppsala, Sweden. Since his time, hundreds of studies have employed regression analysis to explain and predict city temperatures. These predictions are based on independent variables related to urban form and function, land use and cover, city size and population, and regional sky and soil conditions. The results show that CUHI magnitude is predicted reasonably well (at city scale) by independent variables relating to urban land area, city population, and regional weather (cloud amount, wind speed). However, if the scope of the CUHI study is local rather than city, the range of dependent and independent variables can be extended to provide greater understanding of the causal relations. Dependent variables might then include intra-urban temperature differences between pairs of dissimilar LCZ classes (e.g., $\Delta T_{LCZ\ 1-9}$, $\Delta T_{LCZ\ 2-D}$, and so on), while the independent variables will include measures such as sky view factor, mean building height, pervious surface fraction, and thermal admittance of soils. This possibility underscores the basic need for metadata in a CUHI study if one wishes to explain or predict heat island magnitudes or time-space temperature patterns.

The choice of variables to use in a regression analysis relies fundamentally on the metadata that are available, and on the objectives and targeted audiences of the study. Consider the example of a weather office that wishes to adjust urban forecasts of daily minima based on the likely strength of the CUHI: if temperature data have been gathered at stations in and around the city for several years, a simple mathematical function that accounts for meteorology may be suitable, as long as the sites of prediction and observation correspond. For other practical needs such as landscape design or land-use zoning, a set of regression inputs related to the surface structure and cover of the study area should be employed to predict ΔT_{u-r} or $\Delta T_{LCZ\ x-y}$.

While regression analysis is an instructive tool for CUHI investigators, there are notable caveats. First is the instability of regression equations. This means that their predictive power is limited to the places and circumstances under which the values of the independent variables were measured. In our previous example, the equation developed to predict nightly CUHI magnitudes for a local weather office is valid (i.e., stable) only for the city and sites at which the inputs were obtained, which is to say that the regression coefficients and their predictions are not easily transferable to other places (or measurement conditions). This may not be a limitation for CUHI studies intended only for local audiences or applications. If, however, the input data for a regression equation can be gathered from a diverse set of cities, the stability of the equation will increase. One can then begin to extract universal relations (e.g., Eq. 2.13) that require testing in other cities and regions, and that stimulate new and intriguing research questions.

A second caveat is that the mathematical functions of regression analysis are simplifications of reality, meaning that many important factors influencing the dependent variable (ΔT_{u-r}) are deliberately excluded from one's interpretation of the phenomenon (i.e., the heat island). Ultimately, the outcomes of regression are biased to the variables included in the equation. Adding more independent variables is a common response to overcome this bias and to increase the correlation coefficient (r). However, care must be taken not to over-specify by adding independent variables that are unrelated to the physical causes of the CUHI. Furthermore, too many variables can make the equations awkward for practical use, given the demands for data input. A compromise must be met between the required input of the equation and the accuracy of its output. It is therefore necessary that CUHI investigators consider input variables that are *measurable* according to internationally agreed standards, *compatible* with the specific goals of the study, and *relatable* to the physical causes of urban heat islands (i.e., surface energy and water exchanges).

5.4.3   Numerical models

It is always beneficial to link changes in T_a to their causative (energetic) processes (Chapter 2). This can be done in a CUHI study using a simple urban energy balance model such as SUEWS (Section 2.3.3). The input requirements for these models are very modest—in fact, much of the input can be sourced from the metadata files of the measurement sites. The metadata, along with the LCZ parameter values in Table 2.4, will help to drive these simple energetic models. The output will allow users to interpret a measured T_a value (and thus ΔT_{u-r} and ΔT_{LCZ}) as a response to the energy exchange processes operating at the surface (Q^*, Q_H, Q_E, ΔQ_S).

5.4.4   Numerical temperature indicators

Temperature indicators usually measure the number of days above or below a threshold value for a specified time, place, and purpose. The number of annual "frost days," for example, is the sum of all days within a given year and for a specific place having a daily minimum T_a below 0 °C. Other indicators such as heating degree days, cooling degree days, growing degree days, hot days, and tropical nights have different threshold temperatures but similar methods of calculation. Definitions of these indicators will vary with region and user community; if the indicators are used in a CUHI study, clear statements must be given to define the threshold temperatures and operational terms.

Temperature indicators are well suited to CUHI studies because their computed values are known to vary among and within the built and natural environments of a city. In addition, since the indicators are derived from standardized definitions, comparison of their values with other cities becomes possible. This is especially helpful to share common challenges or best practices among cities, for example to mitigate urban heat or adapt to local climate change. Also possible is the delivery of climate information to its relevant users, such as planners, architects, engineers, and health professionals. Cooling degree days, for example, relates the energy demand for building cooling to the local and regional climate; its values differ between urban and rural areas and between compact and open LCZ classes, especially if changes in terrain are involved (elevation, surface relief). This is also true for the number of "hot days" ($T_{max} > 30\,°C$) and "tropical nights" ($T_{min} > 20\,°C$), which relates the local and regional climate to the demand for household energy and water, and to the onset of human heat stress and heat mortality.

While the temperature-energy relation has many important uses (e.g., regional planning; energy and housing policy), one must recognize that it varies with scale and with climatic and nonclimatic factors. For example, at the building scale, the relation changes with the efficiency, or resistance to heat flow, of individual structures—this is determined by architecture, insulating materials, and heating and cooling systems. At the city scale, the relation varies with daily cycles of human activity (e.g., cooking, sleeping, commuting), which consume energy independently of the mean daily air temperature. Finally, when interpreting the values of a temperature indicator, one must always consider the local environment of the measurement sites. This is to ensure that the observed temperatures are representative of the region to be monitored, and that proper explanations are given for the time-space variances in the indicator values (see Section 4.2.3). Furthermore, at least one year of hourly temperature data is needed from a network of climatological stations to study spatial patterns in indicator values.

When combined with other thematic indicators related to environment, economy, health, and housing, temperature-based indicators can inform sustainable and resilient planning at city or neighborhood scale, while improving urban-planning instruments such as building codes, master plans, and zoning schemes. Areas of the city known to experience frequent hot days or tropical nights should be tracked for other numerical indicators relating to shelter, mobility, poverty, unemployment, green space, and air pollution. This information is essential to data-driven policies for climate-sensitive urban design, and it sets a critical benchmark for measuring progress toward healthy and efficient cities.

5.5   Illustrating the data

The results of a CUHI study can be illustrated in many graphical forms, depending on the type and quantity of the data. Here we focus on the use of maps to display heat island results to a diverse group of users. Maps are the most effective way to visually portray the spatial and temporal patterns of the heat island, and to communicate the results intelligibly. The *isothermal* map is the most interpretive format of a CUHI illustration. At city scale, it conveys the "morphology" of the heat island through the configuration of its isolines; at neighborhood scale, it shows the temperature distribution of one or more local areas in the city. While the maps are relatively simple to construct with computer software, they do require a spatially dense array of temperature samples so that the interpolation scheme (e.g., kriging) can draw statistically reasonable lines. In areas of the map where data points are sparse or nonexistent (e.g., rural areas or places of restricted access), the interpolated lines should be removed from the map or else depicted in a distinct line-style (e.g., dotted or dashed; see Fig. 1.4) to indicate uncertainty in their position. Any lines that extend beyond the sampled area are, in effect, climatologically fictional.

Fig. 5.4 for Mexico City—much like Fig. 1.4 for London—represents a classic example of a CUHI isothermal map. Its horizontal patterns for the early morning of 8

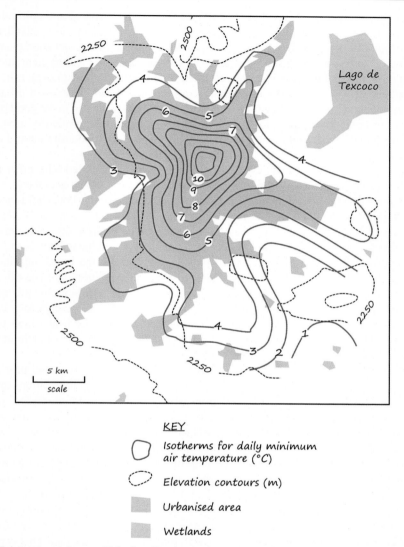

Fig. 5.4 Isothermal map of Mexico City for the early morning of 8 February 1972. Weather conditions: clear and calm.
Redrawn from Jáuregui, E., 1973. The urban climate of Mexico City. Erdkunde 27, 298–307.

February 1972 are based on measurements of minimum T_a made at dozens of climate stations in the city and its environs (Jáuregui, 1973). When viewing the figure, notice how the isotherms do not extend beyond the urban area—this is due to the scarcity of data in the countryside. Inside the urban area, the lines are drawn with confidence and are shown to correspond closely with the physical limits of the city and its areas of greatest density, mainly the city center. Also evident are the effects of surface relief on CUHI morphology, with cooler air from the western and northern hillslopes draining into the urban area.

Therefore, in addition to isolines of equal temperature, physical and cultural features of the landscape that influence T_a should be included in the heat island map. These might be water bodies, built-up areas, transportation routes, elevation changes, and major land uses. Also recommended are the locations of measurement sites. A series of isothermal maps—when placed in chronological order—can depict the correspondence between changes in diurnal, seasonal, and annual patterns of urban air temperature and the underlying form and function of a city. One can make good use of this cartographic information for applied purposes, for example to identify urban hot spots for localized heat mitigation or adaptation measures, or to identify natural areas where plant growth has been modified by the CUHI effect.

Heat island maps are made even more effective when illustrated with other attributes of urban climate (e.g., wind, air pollution), urban form (street geometry), urban metabolism (energy use, population density), social demography (household income, age structure of population), and physical terrain (relief, land cover, soils). Using a Geographic Information System, these spatial data can be layered into environmental maps that highlight the interplay between social and physical characteristics of a city. This approach to urban climate studies (which originates in the industrial cities of Germany) has been used for several decades to communicate urban climate information to land-use planners. The portrayal of this information as urban climate "analysis" and "recommendation" maps for planners defines a popular methodology called Urban Climatic Mapping, or UCMaps (Ng and Ren, 2015). The heat island layers that should be included in a UCMap are critical to the overall transfer of planning and policy information to city departments and health professionals.

Other illustrations to convey the results of a CUHI study include scatter plots, line graphs, and bar charts. The data that can be represented in these illustrations are diverse: they might include functional relations between CUHI magnitude and measures of urban form, city population, and regional meteorology; or frequency distributions of T_a and heating/cooling rates at urban and rural sites. If there is emphasis on small spatial scales, illustrations to convey temperature differences among and within LCZ classes of urban and rural areas can be helpful. These are best displayed using box plots, which are statistical charts that show the spread of data (e.g., temperature) within specified groups or classes (e.g., LCZ). Box plots indicate the median values for these groups, as well as their range (maximum/minimum) and distribution (upper/ lower quartiles). In your work, you might find that the median temperatures for the same LCZ class vary considerably among different locations in the city. This is partly owed to the class locations relative to surrounding LCZs, and to changes in elevation, surface relief, and nearness to waterbodies. Slight variations in the form and/or function of an LCZ class will also contribute to its temperature variability across different areas of the city. Fig. 5.5 presents a series of box plots to illustrate these possibilities, using data from a mobile temperature survey in Vancouver (Canada). In this figure, you will notice that the range of temperature extremes *within the same LCZ class* (i.e., intra-class variability) can exceed the range of median temperatures *between different LCZ classes* (inter-class variability). This is a normal occurrence, and to understand it we must consider that LCZs are defined by their physical uniformity at the *local scale*,

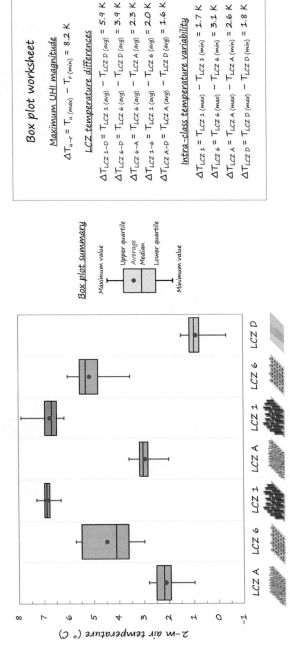

Fig. 5.5 Box plots for 2-m air temperatures in LCZ classes of Vancouver (Canada). Temperatures were measured during a nighttime mobile survey on 4 November 1999. Box plots for repeated LCZ classes represent different locations in the city (see Fig. 5.6 for survey details). Data from T.R. Oke (University of British Columbia).

not the *microscale*. Thus, microscale influences on temperature can arise in every LCZ class. These influences are seen in the spikes and dips of thermal line-graphs obtained from detailed surveys of the CUHI. A clear example of this effect is given in Fig. 5.6, which uses the same T_a data contained in the box plots of Fig. 5.5. If the objective is to investigate heat islands at the local scale, it is important to use spatial averages (or otherwise locally representative sites) to "smooth out" these variations.

Thermal line-graphs are also a simple and effective way to illustrate heat island morphology, especially if the graphs are fitted with a secondary axis to represent changes in land use, surface elevation, building height/density, and LCZ class. The juxtaposition of these climatic and topographic features for an urban-rural area is displayed in Fig. 5.6. With a quick glance at the figure, you can gather a good deal of information about the CUHI effect and its local geographical setting.

Regardless of the graphical method chosen to display the heat island data, one should (if possible) use graph axes that are nondimensional, e.g., magnitude of the CUHI expressed as a fraction of its largest observed value, or the times of observation normalized to the times of local sunset and sunrise. Normalizing the data in this way eases comparison of results with other studies, and allows general relations to emerge from large bodies of work. Furthermore, when communicating the data, one should

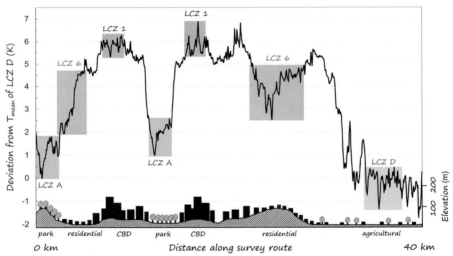

Fig. 5.6 Sectional form of the nocturnal CUHI in Vancouver (Canada) on 4 November 1999. Air temperatures were measured at 4-s intervals along a mobile survey route. Skies were clear and calm during the survey, and all data were adjusted to a common base time of 2200 h. Graph line shows deviations from the mean air temperature of LCZ D (see Fig. 5.5). Good correspondence is observed between temperature deviation and land use, surface elevation, building height/density, and LCZ class. The approximate elevation of the surface is indicated by the hatched area at the base of the cross-section.
Redrawn from Stewart, I.D., Oke, T.R., Krayenhoff, S.E., 2014. Evaluation of the 'local climate zone' scheme using temperature observations and model simulations. Int. J. Climatol. 34, 1062–1080.

always consider the *mean* conditions at urban, suburban, and rural sites, as well as (i) deviations from the mean conditions, (ii) ranges of observed values, and (iii) extreme cases. This information is vital for planning purposes, partly to ensure that physical and social infrastructure (e.g., hot-weather shelters, heat alerts) can meet the needs of vulnerable people during critical warning states, such as extreme hot/cold days or prolonged heatwaves. This makes essential the collection of temperature metadata and the communication of results in the ways that we have outlined in this chapter.

5.6   Concluding remarks

You have now been introduced to the analysis, interpretation, and presentation of data in a CUHI field study. The treatment of these data involves five critical steps. If followed in correct order, the steps ensure that the CUHI measurements are qualified for use in a scientific study of the urban climate, and that they support reliable and compelling outcomes. The following checklist summarizes these key steps.

1. *Compile metadata*: Gather descriptive information on (i) instrumentation, (ii) observing practices, (iii) measurement sites, and (iv) weather conditions during the study period.
2. *Quality control of metadata*: Use the metadata files to identify flawed measurements arising from instrument malfunctions, human errors, and peculiar changes to the study environment.
3. *Stratify metadata*: Sort the metadata according to site characteristics, weather conditions, and observation times and dates. This is crucial for identifying causal factors behind the observed CUHI effects.
4. *Process metadata with CUHI observations*: Use correlation/regression analysis, urban energy balance models, and/or numerical indicators to establish relations between heat island effects, regional weather conditions, and urban form and function.
5. *Illustrate the CUHI observations*: Create isothermal maps to communicate the spatial and temporal patterns of heat islands. Use box plots to display the spread of temperature values within and between data groups, such as LCZs.

References

Aguilar, E., Auer, I., Brunet, M., Peterson, T.C., Wieringa, J., 2003. Guidance on Metadata and Homogenization. WMO Technical Document No. 1186, World Meteorological Organization, Geneva.

Jáuregui, E., 1973. The urban climate of Mexico City. Erdkunde 27, 298–307.

Lowry, W.P., 1977. Empirical estimation of the urban effects on climate: a problem analysis. J. Appl. Meteorol. 16, 129–135.

Ng, E., Ren, C. (Eds.), 2015. The Urban Climatic Map: A Methodology for Sustainable Urban Planning. Routledge, London.

Oke, T.R., 2006a. Initial Guidance to Obtain Representative Meteorological Observations at Urban Sites. IOM Report 81. World Meteorological Organization, Geneva.

Oke, T.R., 2006b. Towards better scientific communication in urban climate. Theor. Appl. Climatol. 84, 179–190.

Runnalls, K.E., Oke, T.R., 2000. Dynamics and controls of the near-surface heat island of Vancouver, British Columbia. Phys. Geogr. 21, 283–304.

Stewart, I.D., 2011. A systematic review and scientific critique of methodology in modern urban heat island literature. Int. J. Climatol. 31, 200–217.

Sundborg, Å., 1951. Climatological studies in Uppsala with special regard to the temperature conditions in the urban area. In: Geographica 22. Geographical Institute of Uppsala, Sweden.

WMO (Ed.), 2008. Guide to Meteorological Instruments and Methods of Observation, seventh ed. World Meteorological Organization, Geneva. WMO-No. 8.

WMO, 2017. International Cloud Atlas. Manual on the Observation of Clouds and Other Meteors, vol. 3 World Meteorological Organization, Geneva. WMO-No. 407. Available at https://cloudatlas.wmo.int/home.html.

Conducting a SUHI study

6

Although the surface and canopy-level urban heat islands (SUHI and CUHI) are clearly linked, the former has historically not received the same attention from the scientific community. This may seem surprising to you given that the causes of the CUHI refer explicitly to (i) the thermal properties of urban fabric that can store heat efficiently; (ii) the lack of vegetation at the surface that limits evaporative cooling; (iii) the geometric arrangements of buildings that restricts the sky view factor (SVF) and surface cooling; and (iv) heat added owing to energy loss from buildings. In short, all of the local controls on the CUHI are fundamentally associated with the character of the urban surface; moreover, most of the heat mitigation and adaptation measures discussed in Chapter 3 are focused on managing the CUHI and SUHI together through manipulation of this surface, for example by increasing vegetative cover and increasing albedo.

The reason for the distinction between CUHI and SUHI studies is simply that, until recently, it was very difficult to observe surface temperature (T_s) over broad and diverse landscapes. Unlike air temperature (T_a), which naturally integrates the effects of a myriad of facets in the vicinity of the thermometer (that is, its source area), T_s is directly linked to the unique exposure and properties of individual facets—even small differences in slope and aspect can result in large thermal disparities (see Fig. 2.16). In complex urban environments characterized by facets with different geometric, radiative, and thermal properties, T_s is variable over very short time and space intervals. Obtaining representative values requires a sampling scheme that can account for the 3D distribution of T_s.

In this chapter, we introduce you to SUHI research practice in the same vein as Chapters 4 and 5 for the CUHI. The emphasis here is on the use, analysis, interpretation, and presentation of satellite-based land surface temperature (LST) data, which is now commonplace.

6.1   Preparing the study

In Chapter 4 we began with three reflective (and related) questions for you to consider: *What is the purpose of your CUHI study? What are the resources available to you?* and *Who is your target audience?* These are equally applicable to planning a SUHI study, as the answers will determine its spatial and temporal coverage, the scale of the observations, the type of analyses, and the communication of results.

The Urban Heat Island. https://doi.org/10.1016/B978-0-12-815017-7.00006-0

6.1.1 Purpose

The SUHI literature can be crudely sorted into (i) field experiments in selected microscale sites, (ii) neighborhood and city-scale investigations, and (iii) comparative studies of multiple cities. Each type of study aims to describe the causes of the SUHI, the geography of T_s (at different scales and on different facets), and the relation between T_s and attributes of surface cover, materials, and geometry. More so than for CUHI studies, those of the SUHI are targeted to urban areas of high temperature (hot spots) and daytime heat extremes, rather than nighttime warming. Depending on the purpose of the study, ancillary databases will be needed to analyze T_s; typically these include the thermal/radiative properties of materials, the water/vegetative surface cover, and the geometry of urban facets (e.g., roof and walls) and units (e.g., streets and courtyards). The study may be completed for research purposes (e.g., a novel environment, or as part of a comprehensive UHI survey, or to evaluate the results of a simulation model); for teaching purposes (e.g., to demonstrate a link between urbanization and urban warming); or for policy purposes (e.g., to identify hot spots, or to monitor the impact of urban landscape change, or to evaluate the effect of heat mitigation plans).

6.1.2 Resources

The design of any SUHI project must take account of available resources (e.g., time, equipment, vehicles, field sites, computers, software). A key consideration is whether you are conducting firsthand observations or using secondary data gathered by other means, such as satellite-based temperatures originating from Earth Observing Systems (EOS). If it is the former, then you will have to consider instrumentation (in situ or remote), platforms (fixed or mobile), and perspectives (ground level or elevated).

6.1.2.1 Surface temperature instruments

The temperature of a surface can be acquired by contact (in situ) using conventional thermometers, or at distance (remotely) using thermal infrared (TIR) sensors. Contact-based measurements require the thermometer to be placed within the fabric, close to the surface yet in thermal equilibrium with its surroundings (cement, tile, brick). Practically, this can only be achieved by using electrical thermometers (such as thermocouples, Section 4.2.2) that are carefully embedded in the selected fabric. If you are making measurements on different facets in the field, you will have to consider how to record the data at multiple locations. While the materials used to make individual temperature sensors are inexpensive, data loggers to interrogate the sensors, calibrate the signal, and store information at frequent intervals are costly. Typically, contact measurements are associated with microclimatic field studies. For example, Offerle et al. (2007) measured T_s by attaching thermocouples to wall surfaces within a city street as part of a detailed study on energy exchanges within the urban canopy. The sensors were matched to the fabric by mixing finely crushed brick with adhesive.

More commonly, T_s is acquired remotely using TIR sensors that measure the long-wave radiation received from a surface within a field of view (FOV), and convert this measurement into a temperature value. The FOV describes the angular properties of the lens through which the longwave radiation enters the instrument (a smaller angle indicates a more restrictive view). The actual area of surface "seen" by the TIR sensor is determined by its FOV and the distance separating the instrument from the surface of interest. In the field, these instruments can be deployed in a variety of ways (Fig. 6.1). Inexpensive instruments that are used to measure the temperature of hot/cold objects will respond to a wide range of temperature values but with large measurement uncertainties ($\pm 2\,°C$) and poor spatial resolution (Fig. 6.1A). By comparison, TIR cameras used for examining heat loss from buildings provide greater resolution but are moderately expensive (Fig. 6.2). Research-level instruments can be very expensive and unwieldy, requiring tripods (for a fixed survey) or stable platforms (for a mobile survey). The instruments have a relatively narrow FOV ($< 45°$) and are pointed at particular surfaces (Fig. 6.3).

Pyrgeometers are micrometeorological instruments that record longwave radiation from a near-hemispheric FOV. They are commonly configured with a series of upward- and downward-facing radiation devices to measure the components of net radiation (Q^*), which is one term in the surface energy balance. These devices can be configured to measure radiation exchanges at specific facets but are more commonly

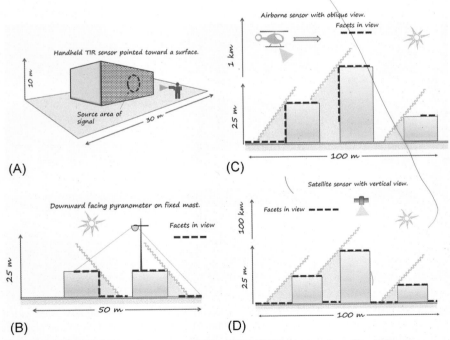

Fig. 6.1 Measuring urban surface temperatures with (TIR) sensors that offer different perspectives of the surface: (A) hand-held sensor pointed by an observer at ground level toward a wall surface of interest; (B) sensor mounted to a mast above the urban canopy layer (UCL), with a circular source area; (C) airborne sensor with an oblique view of vertical and horizontal facets (dashed line); and (D) satellite sensor with a vertical perspective that views horizontal facets (dashed line).

Fig. 6.2 Left: thermal (top) and visible (bottom) images of a partially shaded wall facet, as recorded by a FLIR C2 camera with a measurement uncertainty of ±2 °C. Right: a hand-held reading of the asphalt road temperature.

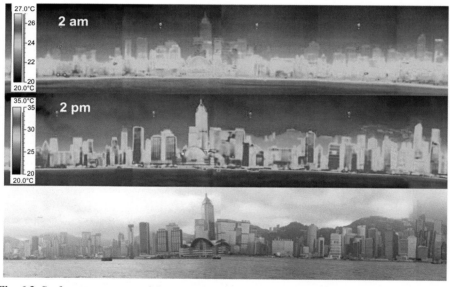

Fig. 6.3 Surface temperature of the Hong Kong skyline and the overlying air as viewed with a ThermaCAM S40 infrared camera on 27 May 2008. The camera is located on a fixed platform with an oblique view. It acquires radiation in the 7.5–13 μm spectral range and has a 1.3 mrad instantaneous field of view (IFOV) and thermal sensitivity of 0.08 K. The images were used for a research project on the ventilation of canopy-level air in a high-density city. The high temperatures of the building walls drive thermal buoyancy, which transfers heat into the overlying air while drawing cooler air into the canopy.
From Y. Li (University of Hong Kong).

placed well above the heights of buildings to record the radiation energy fluxes of the underlying urban surface (Fig. 2.15). The large FOV means that they acquire radiation from a circular source area underneath, and the estimated T_s integrates the contributions of all the facets within view (Fig. 6.1B). Pyrgeometers are expensive (especially when part of a suite of radiation instruments) and require data loggers to interrogate the instrument and store the data.

6.1.2.2 Fixed surveys vs. mobile surveys

Decisions on instruments are directly related to research design, which is often determined by the location of fixed sites and/or the configuration of mobile surveys. Sites for a fixed survey may be selected to represent a typical urban environment (such as a street canyon or a neighborhood type), or to evaluate a specific thermal effect (such as heat stress in a public gathering place). Access and placement for the instruments are key considerations to ensure that the surface(s) of interest can be measured for the duration of the study period (Fig. 6.3). Getting permission to erect an instrumented mast that provides aerial views can be extremely difficult—you may instead wish to use existing field sites where suitable platforms are already in place.

Ground-based mobile surveys of T_s usually involve TIR sensors attached to vehicles for continuous recording of surfaces in the FOV. In an early experimental study on the formation of the nighttime SUHI and its relation to SVF in the urban canopy layer (UCL), Eliasson (1990/91) used a car-mounted TIR instrument to observe road temperatures, which were recorded every 5 m along a transect route. If you wish to employ a similar methodology, the challenge is to identify an appropriate route that can be completed within a useful timeframe, so that the urban controls can be separated from climatic factors associated with daytime warming or nighttime cooling (see Fig. 4.6).

The most common approach to gathering urban T_s is to use TIR sensors on airborne platforms that follow predetermined flight paths. The platform height and instrument attributes (FOV and direction) determine the size and character of the urban surface in view, that is, the distribution of walls, roofs, trees, etc., that compose the radiation signal. Depending on the time of day and the survey configuration, different facets may be in shadow or shade and the contributions will change with the passage of clouds, the relative movement of the Sun, and changes to the flight path. The result is an extremely heterogenous and dynamic pattern of surface temperatures. At night, under clear and calm conditions, facets cool down at different rates based mostly on longwave radiation loss (controlled by the sky view factor), and the temperature responds to the substrate heat available for withdrawal.

Airborne surveys are expensive to conduct and are constrained by weather conditions. For this reason, it is difficult to obtain repeated observations of the same place. Moreover, obtaining permission to fly over a city at a given height may not be possible. An alternative approach is to use existing thermal data that have been collected for other purposes, e.g., to identify building energy loss in cool climates where space heating is required. These data are usually gathered in winter months and in clear, calm conditions when heat loss is maximized. Fig. 6.4 illustrates this type of work, which was completed primarily to identify poorly insulated buildings; the figure also

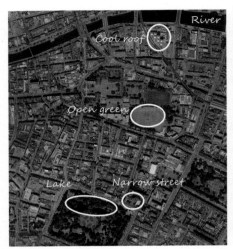

Fig. 6.4 Visible and thermal images of Dublin (Ireland). The thermal image (right) was taken on a cold and clear night between 0120 and 0400h using a CS-LW640 aerial digital camera. The measurements were designed to record relative surface temperature (dark red/cold to yellow/warm) for assessing building energy loss, and must be calibrated against a known surface temperature to estimate T_s. Note the correspondence between relative surface temperature and surface properties (structure—SVF; fabric—roof and street; cover—vegetation/water).

illustrates variations in T_s that correspond to different surface types (paving, vegetation, water). Data of this kind may suit your needs (in terms of scale and weather conditions) and represent a cost-effective solution, but you will not have control over the instruments or the survey timing, which will seriously limit your ability to simultaneously measure ground-level T_a and T_s. Furthermore, the survey may not provide a suitable reference against which to judge the urban effect.

6.1.2.3   Satellite-derived land surface temperature (LST) data

The most commonly used TIR datasets to assess the SUHI are obtained from sensors placed on satellite platforms to record the Earth's LST (Fig. 6.1D). Many LST datasets are free to use and available via websites (e.g., U.S. Geological Survey https://earthexplorer.usgs.gov/). Moreover, the data are geographically referenced and quality controlled, and with near-global coverage; however, for places of interest, data characteristics will vary with the satellite mission (its orbital route and frequency) and the properties of the TIR sensor. Satellite missions often include multi-spectral sensors that record LST-related information about the landscape, such as the health and coverage of its vegetation. Despite the conveniences of LST data, they do have limitations based on the sample acquired and the sensor's view of the urban surface. When working with satellite data, you will require computer resources to manage very large datasets, along with software and operational skills to analyze LST data. A typical Landsat scene (see Fig. 2.19) covering an area of 170×180 km will contain about 250MB of digital information.

6.1.3 Target audience

In previous chapters, we divided the audience of your heat island work into two overlapping groups: scientists and/or policy makers. The former group is interested in the originality of the work, while the latter needs evidence to support the development of UHI policies for managing the urban temperature effect. Originality can be judged in many ways but is usually related to the existing literature in the field, and to identifiable gaps where little research exists and thus the proposed research is novel. The only way to discover what is original is to immerse yourself in the published literature, as this provides a basis for identifying appropriate avenues and audiences for your work—the research illustrated in Fig. 6.3 is an example. For policy-based work, the emphasis shifts from scientific originality to the applicability of the findings. In some cases, the scope and objectives are clear at the outset and the methods and topics may be well established; however, the audience may wish to have results that are presented differently, often in terms of priorities for mitigation actions based on public health or economic costs and benefits—the work illustrated by Fig. 6.4 is an example. The academic and policy audiences may often overlap, particularly because research is lacking on the application and evaluation of scientific knowledge on urban heat islands.

6.2 Planning the study

The design of a SUHI study should reflect its purpose and the availability of resources. A microscale field experiment is typically a study of short duration and marked by intensive deployment of instruments in selected settings, which are chosen for their spatial representativeness (e.g., street canyons) or for a targeted phenomenon (e.g., heat stress in the outdoor environment). In these studies, T_s is usually just one variable of many that are measured, sometimes as a part of a larger field study that may not be designed as a conventional UHI investigation (i.e., as a comparison of urban field sites with natural or rural sites).

In theory, a SUHI study would include the "complete" surface temperature (i.e., T_s at all facets constituting the 3D urban surface) and its diurnal response to changes in weather and seasons. In practice, however, SUHI studies can only include "incomplete" surface temperatures over greatly reduced subsets of facets, and with repeated measurements over space and time. Subsequent geographical analysis of those temperatures will link the spatial patterns to the distribution of the properties of the urbanized landscape (structure, cover, fabric, and metabolism). Temporal analysis will track changes in the SUHI over time and in correspondence to shifts in the climate and the extent and intensity of urbanization. Multi-city studies have elements that are similar to single-city studies, but seek to identify the intrinsic (city) and extrinsic (geographical context) factors that result in inter-urban SUHI differences. In these cases, the SUHI is defined as a measured difference between urban T_s and background T_s, the latter being the natural landscape in the vicinity of the city and that represents the pre-urban condition.

6.2.1   Methods of observation

To measure surface temperature on contact, one must place thermometers on, or very close to, the surface of the selected fabric. The observational challenge in this arrangement is to select sites and fabrics that are representative of the urban landscape. For this reason, contact-based measurements are rarely employed in the field, with the exception of laboratory-like settings in which the variabilities of fabric and geometry are limited. One example is the testing of materials in outdoor conditions where different fabrics are arranged to have the same exposure to weather. Each sample of fabric has been modified (to include a thermometer) so that its response to weather over time can be observed. These types of experiments can provide valuable insights into the potential for novel fabrics and coatings to limit surface warming and mitigate excessive heat, but are not used to directly assess the SUHI.

Noncontact techniques require TIR sensors to record the longwave radiation emitted from a surface, and to convert this signal into a temperature value (Fig. 6.2). This conversion is based on Planck's Law, which states that the black-body emission at any wavelength is a function of its surface temperature (T_s) and emissivity (ε_s):

$$L\uparrow_\lambda = f\left(T_s^4 \varepsilon_s\right)$$ (6.1)

The term *black body* refers to an object that absorbs all radiation that it intercepts, and reflects none of that radiation (that is, a perfect emitter, $\varepsilon_s = 1$). In the longwave spectrum, most natural objects are close to behaving as perfect emitters with emissivity values of > 0.85 (see Table 2.1).

TIR sensors have a FOV that is measured as a solid angle, which describes a cone extending from the instrument in the direction of the object of interest. The size of the surface area from which the radiation originates depends on the FOV and the distance between the instrument and the surface. Unlike an air thermometer, which is exposed to the flow of air from all directions and therefore records the contributions of a great many facet of facets, a surface thermometer is directionally exposed, meaning it is positioned to see a surface of interest (Fig. 6.1). If the instrument is perpendicular to a flat surface, the area seen is circular in shape. Under the assumption that the surface of interest emits equally in all directions (i.e., isotropic emission), then the radiation received by the instrument does not vary with its orientation relative to the surface. TIR cameras use an array of detectors to decompose the radiation in a FOV into picture elements (pixels), each of which has a spatial resolution that depends on the size of the FOV, the size of the sensor array, and the distance separating the instrument from the surface of interest.

TIR sensors record longwave radiation originating from different sources and processes, namely (i) emitted by the surface of interest, (ii) reflected from the surface of interest, and (iii) emitted by the intervening atmosphere. When the distance between the instrument and the surface is small (i.e., the instrument is at or near ground level and directed toward a nearby surface), atmospheric interference is minimal; however, at greater distances (e.g., mounted to an aircraft or satellite), this interference must be removed. This is substantially accomplished by ensuring that the sensor responds to wavelengths between 8 and 14 μm (the atmospheric window), in which the atmosphere is a poor absorber (and emitter). Even still, for any appreciable distance between the instrument and the surface, the radiation received must be further corrected to remove the atmospheric effect, especially when humidity and aerosol content is high. What remains is longwave radiation received from the surface in the FOV.

If we assume that this surface is a perfect emitter (or black body, $\varepsilon_s = 1$), then there is no reflection and the estimated temperature is known as the brightness temperature (T_b), which is always higher than the actual T_s. Most natural and manufactured materials have ε_s values of between 0.80 and 0.95, meaning that 5–20% of $L\downarrow$ on the surface in the FOV is reflected radiation and must be discounted in any signal. Manufactured materials have consistent ε_s values, but those of natural surfaces change with soil water content and vegetation growth patterns. In addition, some commonly manufactured fabrics, such as corrugated iron (which is used as a roofing material in many parts of the world), have very low ε_s values that make them poor emitters and excellent reflectors (Table 2.1). Simple hand-held infrared thermometers allow users to adjust the emissivity values based on the properties of a nearby uniform surface (e.g., wall or pavement). For more distant surfaces, the FOV will capture radiation from a mix of surface types with diverse ε_s values. In this case, selecting a single emissivity value will result in errors when estimating T_s.

6.2.2   Satellite observations

A major source of thermal information for SUHI studies is the satellite-based TIR sensor, which has been associated with Earth Observing Systems (EOS) since the 1970s. EOS satellites can be geostationary (with a fixed view of a portion of the Earth's surface) or orbital (with a constantly changing view along a set route). The latter are usually polar orbiting and synchronized with the Sun so as to pass over the Earth's Equator at the same time every day. Global coverage is acquired as the Earth rotates, so that relative to the surface, the satellite's FOV outlines a corridor (swath) across the Earth (Fig. 6.5). By repeating the same route, the satellite eventually acquires near-global coverage while passing over the same place at regular intervals. The width of the swath is determined by the FOV of the instrument at its nadir and its side-to-side exposure, which is controlled by mirrors. Several EOS will have TIR sensors in addition to other sensors that record reflected shortwave radiation, often in specific bands (that is, multi-spectral sensors, MSS).

The TIR sensors are used to derive LST and thus the temperatures of urban surfaces. Satellite TIR/LST data vary by time period, resolution, accuracy, and availability (Table 6.1). Two of the most commonly used sensors for SUHI studies are Landsat and the Moderate Resolution Imaging Spectroradiometer (MODIS). MODIS is sensitive in 36 spectral bands ranging in wavelength from 0.4 to 14.4 μm. It records thermal information in two narrow thermal bands centered on 11 and 12 μm. The same instrument has been fixed to two satellite platforms (Aqua and Terra) that follow sun-synchronous, near-polar orbits, passing overhead at 1330 and 0130 h (Aqua) and at 1030 h and 2230 h (Terra) local time. The FOV is 10 km wide along its track at nadir but the cross-track swath is 2330 km wide, allowing the Earth's surface to be scanned every 1–2 days. The data gathered are quality controlled and made available as a global product at 1-km resolution, representing monthly and 8-day averages.

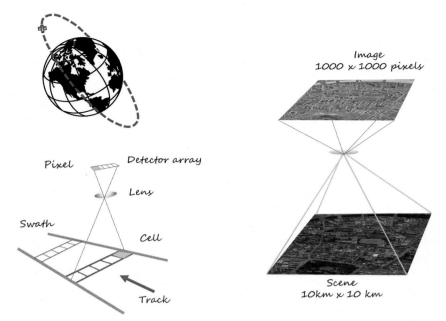

Fig. 6.5 Operation of a polar-orbiting satellite and the radiation information that it gathers from a swath of cells on the ground below. The thermal signal is focused onto detector arrays where each device corresponds to a pixel. On the right, a scene covering an area of $100\,km^2$ is depicted as an image of 1,000,000 pixels, corresponding to a spatial resolution of 10 m.

Table 6.1 Specifications for satellite TIR/LST data by sensor type.

Sensor	Satellite	Spatial resolution	Orbital frequency	TIR spectral bands (μm)	Time of image acquisition
Landsat	Landsat 4–5	120 m	16 days	10.4–12.5	1000 h
	Landsat 7	60 m		10.4–12.5	
	Landsat 8	100 m		10.60–11.19	
				11.50–12.51	
MODIS[a]	Aqua	1 km	Twice daily	10.78–11.28 11.77–12.27	1330 h 0130 h
MODIS	Terra	1 km	Twice daily	10.78–11.28 11.77–12.27	1030 h 2230 h
ASTER[b]	Terra	90 m	Twice daily	8.125–8.475 8.475–8.825 8.925–9.275 10.25–10.95 10.95–11.65	Request only
AVHRR[c]	Multiple NOAA satellites	1.1 km	Twice daily	10.3–11.3 11.5–12.5	Morning and afternoon

[a] Moderate Resolution Imaging Spectroradiometer.
[b] Advanced Spaceborne Thermal Emission and Reflection Radiometer.
[c] Advanced Very High Resolution Radiometer.

Table 6.2 The nine spectral bands for the Operational Land Imager (OLI) and two spectral bands of the Thermal Infrared Sensor (TIRS) on board Landsat 8, launched in 2013.

Landsat 8 Operational Land Imager (OLI) and Thermal Infrared Sensor (TIRS)		
Bands	**Wavelength range λ (μm)**	**Resolution (m)**
Ultra Blue	0.435–0.451	30
Blue	0.452–0.512	30
Green	0.533–0.590	30
Red	0.636–0.673	30
NIR	0.851–0.879	30
SWIR 1	1.566–1.651	30
SWIR 2	2.107–2.294	30
Panchromatic	0.503–0.676	15
Cirrus	1.363–1.384	30
Thermal 1	10.60–11.19	100
Thermal 2	11.50–12.51	100

Source: U.S. Geological Survey (2019).

Landsat is the best known of the EOS satellites. It scans the Earth's surface every 16 days, passing overhead at approximately 1000 h. Landsat 4 was the first in the series to record in the thermal band at a resolution of 120 m (the other radiation bands are recorded at 30 m). The width of its swath is 185 km. Subsequent Landsat missions have gathered information in much the same way, following the same path so that there is an extensive back-catalogue of images. The most recent mission (Landsat 8) gathers data in 11 different bands: thermal data are gathered at a resolution of 100 m, panchromatic at 15 m, and MSS at 30 m (Table 6.2). The thermal data are resampled to 30 m to match the MSS data. The two thermal bands are used to obtain a more precise assessment of surface emissions by removing the atmospheric contribution.

Landsat and MODIS together convey the important choices to be made at the outset of a satellite-based SUHI investigation. These include the orbital frequency of the satellite, the time of image acquisition, and the spatial resolution of the data (Table 6.1). MODIS can provide information on the SUHI of a large, extensive city at a relatively crude resolution (1 km), and it can do so quite frequently (every 1–2 days). Moreover, MODIS can acquire daytime and nighttime images. Landsat, in contrast, can provide more detailed thermal information (100–120 m) but only very infrequently (every 16 days), and the information is available publicly only for daytime. Lastly, the availability of useful thermal data from any satellite system relies on the absence of cloud cover.

6.3   Important considerations

There are several important considerations for you to reflect upon when using TIR data to investigate the SUHI. Here we divide these considerations into three related issues: scale and perspective; sampling; and defining SUHI magnitude.

6.3.1   Scale and perspective

First, consider the information received by an instrument with a vertical perspective over a landscape of heterogeneous cover (grass, paving, water, forest, etc.). When sufficiently close to the ground, the instrument will see an area that contains one land-cover type and the estimated T_s can be attributed to that type. As the instrument is raised from the ground, the area seen by the instrument increases (the spatial resolution becomes coarser) and radiation is acquired from a number of land-cover types. The contribution of each type will depend on its longwave emission ($L\uparrow$) and the proportion of the sensor's FOV that it occupies (Fig. 6.6).

Next, consider a landscape that has the same cover but is corrugated into facets which are either vertical or horizontal (Fig. 6.1). If the instrument is overhead and has a perpendicular view, the bulk of the signal will be acquired from the horizontal surfaces; if the instrument is tilted, it will see more of the vertical surfaces. During the daytime, these surfaces will experience large variation in temperature, which is a response to the geometric relations among the facets, the location of the Sun, and shadowing by obstacles (Fig. 6.3). At night, each facet will cool at a different rate subject to its SVF and the available energy stored during the daytime (Fig. 6.4). Depending on the mixture of facets within the FOV, a different signal is acquired and a different T_s is estimated. Critically, there is no perspective from which the complete surface temperature can be acquired in this scenario, as some facets will always be hidden from view and thus their surface temperatures are not observable.

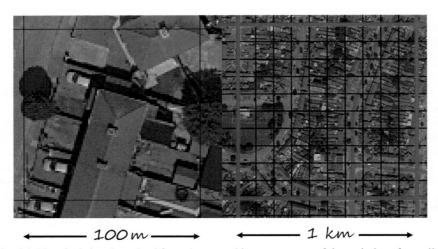

$\longleftarrow$ 100 m $\longrightarrow$ $\longleftarrow$ 1 km $\longrightarrow$

Fig. 6.6 The pixel signal received from the ground is an aggregate of the emissions from all facets in the cell. This includes radiation from roofs, gardens, tree tops, cars, etc. The larger the cell, the greater the number of contributing facets and the range of facet types. The signal received ideally corresponds to a relatively homogenous landscape, so that its temperature can be attributed to particular facets, or common combinations of facets, associated with blocks and neighborhoods.

Both of these considerations are relevant to temperature studies of the urban landscape, which comprises a hierarchy of features (facets, buildings, streets, blocks, neighborhoods) that correspond with variations in land cover, structure, and fabric. Information that can be gleaned from TIR instruments then depends on their spatial resolution and perspective. From the preceding examples, MODIS can capture information on large neighborhoods ($> 1 \, km^2$), while Landsat can acquire information on blocks ($> 100 \, m$). Both instruments provide mostly near-vertical views of the surface. As a result, the contributions of horizontal surfaces (streets, roofs, ground cover, tree canopy tops) are oversampled at the expense of vertical surfaces.

6.3.2 Sampling

Every facet has a temperature signal associated with its surface energy balance, which is an outcome of its unique geometry and material properties, and its relation to the wider urban environment. It is not possible to measure T_s of every facet at urban scales for a given moment in time, so inevitably a sampling strategy is needed. Decisions on how, where, and when to measure (or to acquire) urban T_s are influenced by the purpose of the study and the definition of SUHI.

If you have decided to use satellite LST data, then many of the operational decisions have been made for you and the available data will depend mostly on the weather. Satellite-based sensors can provide a near-complete 2D assessment of T_s across the urban landscape. However, the view of these sensors is dominated by horizontal surfaces. The effects of urban structure are apparent in areas of shade, and in the reduced albedo (increased emissivity) associated with multiple reflection (horizon screening) in the UCL. This creates biases in satellite thermal imagery that will be greatest in densely built parts of the city, where buildings are tall and closely spaced, and where roof surfaces occupy a large portion of the top-down view. It is important that you recognize these biases, especially when linking maps of the SUHI to those of the CUHI, which is based on T_a within the canopy layer (roof surfaces are not part of this layer and their influence on T_a in the UCL is not direct).

The choice of satellite data to use in a SUHI study also affects temporal sampling, which is limited by the orbital frequency and time of passage of the satellite. LST data from satellites are available only under clear-sky conditions; if freely available high-resolution Landsat data are needed, the data are further restricted to one time of the day. It is possible to combine information from several satellites to estimate the diurnal variation in LST, but this will come at the expense of spatial accuracy. The great advantage of satellite data is that they are gathered consistently over several years and are subjected to well-established quality controls. This potentially allows for LST comparisons over time and with seasonal change.

Sampling at the neighborhood and even the city scale can be done with airborne TIR sensors. Depending on the aircraft flight path, the FOV, and the perspective of the TIR instrument, data will be acquired from a myriad of surface types and facets that must be mapped to the urban landscape. During the daytime, the measurement of T_s will depend very much on the path of the flight in relation to the distribution of direct solar radiation at the surface. Flying from north to south in the Northern Hemisphere at noon will oversample from shaded surfaces, while flying in the opposite direction will oversample from sunlit surfaces. It is therefore inevitable that the urban surface will have different temperatures when viewed from different perspectives, a phenomenon referred to as thermal anisotropy. Ideally, a series of flights can be completed within a short time-frame, so that T_s values acquired from multiple perspectives can be measured within a period when the background weather is stable and the geometry of solar radiation has not changed appreciably. During the night, when variation in T_s is a great deal more muted than during the day, the measurement period can be longer. The advantage of this approach is that a degree of control is gained over the timing and configuration of the flight path to reflect the purpose of the study. To match this work with a traditional mobile survey of the CUHI, the path would form a cross-section of the city while ensuring that different land covers are measured.

Pyrgeometers facing toward the urban surface have a large FOV and can be positioned to gather thermal information from an area chosen to represent a neighborhood type. The instrument will receive longwave radiation from a circular source area, the diameter of which varies with its height above the ground: at 20 m, it will receive 90% of its radiation from an area with a 120-m diameter at ground level. However, since the urban surface is 3D, the circular area will be "draped" over the buildings and trees within the source area, allowing some facets to be seen while others are hidden. The question is whether or not the instrument records an "appropriate mix of climatically active surfaces" (Oke, 2006). In other words, is the observation representative of the neighborhood?

Microscale studies require TIR instruments (including thermometers, cameras, and pyrgeometers) to be directed at particular surfaces. A sampling strategy for a microclimatic SUHI study would identify urban configurations (such as city streets, plazas, or courtyards) that are common so that the results can be transferred from place to place. For a case-study SUHI, however, the sites of observation might be chosen for other reasons, for example to assess outdoor heat stress on occupants. In this case, the site choice may be based on persistent "hot spots" identified in large-scale thermal surveys, or on urban design projects to create comfortable outdoor settings. For thermal comfort studies, ancillary information is needed on the radiation environment and on ambient airflow, temperature, and humidity (Section 3.2). T_s data are necessary to assess the mean radiant temperature (T_{MRT}), that is, the equivalent temperature of an enclosure that generates the same radiation loading on the body as that received from diverse sources. In these types of study, globe thermometers can support an analysis of the microscale setting, while simple hand-held thermometers can gather basic thermal information on particular facets.

A good example of a SUHI study using multiple approaches to temperature sampling is the urban boundary layer (UBL) study of Toulouse (France) by Masson et al. (2008) (Fig. 6.7). In that study, TIR measurements were obtained from aircraft transects at constant heights (150 m, 1100 m) to observe the development of the UBL in summer and winter, and during morning, afternoon, and nighttime periods. Alongside other instruments, two TIR cameras with backward inclinations of 10° and 50° were used to acquire T_s at resolutions ranging from 1.5 m to 6 m. To retrieve T_s from the TIR measurements, vertical soundings of T_a and humidity were needed to estimate and

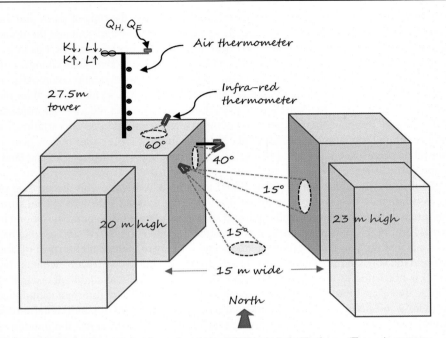

Fig. 6.7 Field site for an urban boundary layer (UBL) study in Toulouse (France), as part of the CAPITOUL project. This component of the project focused on radiation and surface temperature measurements (using Everest TIR instruments) made on the roof, walls, and street.
Redrawn from Masson, V., Gomes, L., Pigeon, G., Liousse, C., Pont, V., Lagouarde, J.P., Voogt, J., Salmond, J., Oke, T.R., Hidalgo, J., Legain, D., 2008. The Canopy and Aerosol Particles Interactions in TOulouse Urban Layer (CAPITOUL) experiment. Meteorol. Atmos. Phys. 102, 135–157.

remove atmospheric effects; additionally, "ground truth" surface-temperature measurements were obtained using infrared thermometers focused on streets, walls, and some roofs on several blocks in the city center. It is clear in this example that designing and conducting a bespoke SUHI study of this scope and at this scale is a substantial undertaking which is not possible without considerable resources.

6.3.3 Defining SUHI magnitude

All UHI studies evaluate the magnitude and timing of the heat island by comparing temperatures made at urban sites with those at benchmark sites. In many such studies, the benchmark is located in a natural area—often outside the urban footprint—to represent the temperature conditions if there were no city in existence. For the CUHI, an official weather station in a rural area often serves as the benchmark, while also providing regional weather data for the study period. Other benchmark sites can be used, but only if the appropriate metadata can be acquired (see Section 4.2.3).

For the SUHI, the choice of a benchmark site depends on the type of study and the system of observation. At the city scale, T_s of the natural landscape outside the urban area is appropriate. This landscape may host a mixture of land covers such as grass, crop, forest, and water, or bare rock, soil, and sand. These land covers can change with the seasons, as plants grow, ground is plowed, trees are harvested, and so on. Also, it is possible that the natural landscape occupies higher (or lower) elevations than the urbanized landscape. The challenge in all such cases is to select the most appropriate site to use as the benchmark after accounting for temporal and spatial variations in land cover and topography (Imhoff et al., 2010). Naturally, large water bodies next to coastal or shoreline cities should be excluded, as these cannot substitute for the landscape in the absence of the city.

Ultimately, the approach to define SUHI magnitude will vary with the method of its assessment. If the study is an airborne survey, the transect path will extend from the urban area to the natural landscape and should be planned to capture variations in urban form in the context of the selected weather. For satellite SUHI studies, one has no control of the observation time and the scene is likely to include all (or most of) the city and the surrounding area. Identifying the limits of a given city (that is, what is "urban") is not straightforward, however. While there are widely available datasets that outline the urban footprint at global and regional scales, each uses different techniques so there is no definitive dataset. Information may be acquired for individual cities, but the data must include the built-up area under study rather than just the municipal boundaries, which may not correspond with the urban extent. It is also possible to use information from satellite multi-spectral sensors to identify impervious or vegetative cover in the study area. Once the urban footprint is defined and unsuitable land covers are filtered out, the potential benchmark information can be obtained from the pixels in the adjacent area.

The effects of topographic context on evaluating the SUHI magnitude can be seen in Fig. 6.8, which presents satellite-observed T_b (i.e., not corrected for surface emissivity) at night (2200 h) and day (1000 h) for the city of Be'er Sheva in the Negev Desert (Israel). During the night, the city appears as a warm area with warmer lines (roads) extending outward. The expected nighttime SUHI emerges as the urban surface cools more slowly than the surrounding arid landscape. During the day, however, the upper layers of the arid land warm up quickly (owing to its low thermal conductivity and heat capacity; see Fig. 2.8), while the city warms slowly, resulting in a reverse SUHI, or "cool island." Outside the city are daytime cool areas that consist of irrigated fields that evaporate freely. In essence, the magnitude and timing of the SUHI are strongly influenced by the nonurban areas chosen as the benchmark.

For neighborhood-scale studies, depending on their purpose, the benchmark site could be a paired urban site chosen to represent a distinct land-use/land-cover (LULC) type. Unfortunately, there is no internationally accepted LULC categorization to support SUHI studies, but the LCZ classification (Section 2.2.3) is a useful starting point. The scheme includes information on the vegetative, impervious, and building surface fractions of urban and rural landscapes, which are known to be energetically linked to variations in T_s and T_a. If LCZ classes are to be used in a SUHI study, the heat island magnitude is then obtained as $\Delta T_{s\,(LCZ\,x\text{-}y)}$, with the statistics of T_s in each LCZ class being compared with those in another.

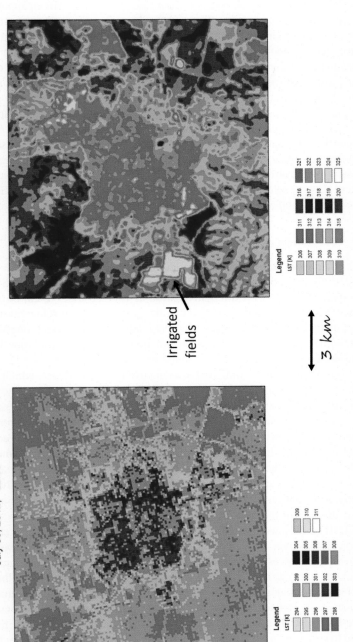

Fig. 6.8 Surface temperature of Be'er Sheva, located in the Negev Desert (Israel). The nighttime image is from the ASTER satellite and the daytime image is from Landsat.
From E. Erell (Ben-Gurion University of the Negev).

At the microscale, SUHI magnitude is obtained as the temperature difference between two or more sites that are relevant to the purpose of the study. For an applied study, e.g., to determine the effect of high-reflectivity (cool) coatings on T_s, the benchmark could be a nearby vegetated surface or a conventional urban surface such as untreated asphalt. In this case, the only requirement is that the instrument setup is identical for each site and that weather conditions are the same. The influence of advection can be minimized by selecting low wind speeds and using plots large enough to ensure that the area observed by the TIR instrument is far from the edges of each plot. For a heat stress study, the objective may be to redesign a public space, in which case the urban temperature effect can be evaluated by comparing T_s before and after treatment at the same site, or by comparing two urban sites that are substantially the same except for one distinguishing property (e.g., different fabrics but same structure). More generally, using a large green park as a benchmark would also suit this purpose. The onus is on you to state the purpose of your study and how the selected sites correspond. The challenge then is to confirm, to the extent possible, that all extraneous circumstances (especially weather) are being managed.

6.3.4   Weather controls

Weather data from stations within the study area are needed to analyze and interpret the SUHI observations (see Section 2.2). Creating a climatology of the SUHI requires you to gather T_s data over a lengthy period of time so that typical diurnal and seasonal variations and likely extremes can be captured. Meteorological data on cloud and precipitation are also relevant: cloud regulates shortwave and longwave radiation receipt, while precipitation changes the radiative and thermal properties of the surface. It is important to keep in mind that the natural landscape will respond to precipitation events over a much longer time period than paved urban surfaces, which are designed to shed water. If you are completing an extensive city-scale SUHI study, the meteorological data should therefore be available for a period beforehand—these data will be needed anyway if you intend to use a surface energy balance model for comparison (see Section 5.4.3), which requires "run-up" time to respond to initial conditions before simulations can be performed. In addition to precipitation, other meteorological variables of importance to the SUHI are wind speed, air temperature, and humidity. These can be used to estimate the magnitude of the surface-air convective exchanges. Similarly, information on global solar radiation $(K\downarrow)$, or even the number of sunshine hours, is helpful for interpreting the observed T_s.

In CUHI studies, weather data are used to discriminate values of ΔT_{u-r} (or $\Delta T_{LCZ\,x-y}$) based on cloud cover and wind, for example (see Eq. 5.1). However, weather conditions do not greatly restrict the assessment of CUHI, whereas SUHI cannot be assessed to any extent when a cloud layer exists between the TIR instrument and the Earth's surface. This puts a major limitation on the use of satellite data for cities that are located in climates with frequent cloud cover, whether partial or complete. Consequently, all satellite-based SUHI assessments are conducted in clear-sky weather, making it impossible to construct a SUHI climatology from this source of data alone. It may be the case that the clearest weather occurs in the winter months so that the available scenes are biased toward a period when little natural energy is available from the Sun. If the purpose of the study is to examine the conditions in which the SUHI is maximized, or simply to examine those times when the daytime or nighttime heat island poses a heat hazard, then the limitation to clear skies may not be a concern.

6.3.5 Processing

TIR data will need to undergo a number of processing steps before they can be used in a SUHI study. The steps include geometric and radiometric corrections to acquire and geolocate T_s values. If your study uses detailed TIR data from an airborne sensor, and the distance from the sensor to the urban surface is large (> 200 m), then you must obtain information on the vertical temperature and humidity structure of the atmosphere (soundings) to estimate its contribution to the TIR signal. Atmospheric soundings are taken at select meteorological stations twice daily, but may also be measured as part of a field study or obtained from numerical weather model simulations. Once the atmospheric signal is removed from the TIR data, T_b can be calculated for the surface(s) in view and T_s is estimated from the surface emissivity. Mapping T_s onto the 3D urban surface must account for the orientation of the instrument, its FOV, and the geometry of the urban surface. Nearby meteorological data (soundings) and details of the urban surface (structure, cover, fabric) appropriate to the spatial scale of the TIR data should also be retrieved. Details on urban morphology (i.e., footprints and heights of buildings) along a flight path can be used to assess the contributions of vertical and horizontal surfaces (sunlit and shaded) to the observed temperatures, relative to the distribution of these surfaces in the study area.

For satellite data, much of the processing may already have been completed within the EOS program. For example, the publicly available Landsat 8 thermal data consist of a matrix of pixels that compose a scene of approximately 170 km $\times 183$ km. Each pixel stores a digital number (DN) that can be converted into top-of-atmosphere near-infrared (longwave) radiation, from which the surface brightness temperature (T_b) is derived using coefficients that are provided in the metadata files accompanying each scene. The metadata also include information on cloud cover and data quality that can help one to select scenes for study. Ideally, you should assemble a dataset that is cloud-free (or nearly so) over the study area. It is also possible that the city of interest spans more than one satellite scene, so that a new mosaicked image can be created by merging adjacent scenes. If the scenes have been acquired on adjoining swaths, the imagery will originate from different days (at least 16 days apart in the case of Landsat). For studies designed to compare the SUHI in a number of cities, the same processing steps must be repeated. When assembling a time series to examine the seasonal variation of the SUHI or its development over time, several years of observations are needed to acquire a sufficiently long dataset. For Landsat on a 16-day cycle, there are 23 opportunities during the year to acquire data. Some data have been processed by the EOS program to provide global coverage, which may be suited to certain SUHI studies. For example, the MODIS product MYD11A2 is obtained as an 8-day LST for an emissivity (ε) average at a resolution of 1 km. The data include daytime and nighttime T_s bands along with associated quality control assessments, observation times, view zenith angles, and clear-sky coverages.

6.3.6 Metadata

Metadata describe information about the T_s data in a SUHI study, including details on the instrumentation and its deployment, the times and dates of the measurement program, and the associated weather. There are no guidelines on the provision of SUHI metadata that are equivalent to those of the WMO on meteorological observations, which outline the types of instrumentation and their exposure. Nevertheless, the intent of the WMO guidelines is clear: it is to establish a set of principles that allow the target audience to trust the data and their wider applicability, and/or allow workers to reproduce the study at other places or times. Our discussion of metadata about instrumentation, observation practice, site placement, and regional weather for CUHI studies (Section 5.1) is applicable here, but the metadata must be tailored to the circumstances of a SUHI study.

TIR metadata should provide the characteristics of the instrument (FOV and radiometric precision) and its deployment (surfaces of interest, 3D position and orientation, and observation schedules) (see Fig. 6.7). If the instrument is fixed to a mobile platform, details of its exposure, the spatial and temporal resolution of data acquisition, and the survey route are needed. Pyrgeometers, in particular, have a large FOV for observing extensive (circular) source areas, based on their height placement. In a typical configuration, pyrgeometers are part of an energy balance measurement system located well above the urban surface, so that the instruments are exposed to a sample of facets that are representative of the selected neighborhood. The supporting metadata should include a map of the surface cover within the source area.

Metadata that normally accompany a satellite thermal image from an EOS mission are comprehensive and provide a useful model. For example, Landsat 8 scenes are graded as "Level 1" if they have been corrected for distortions arising due to the instrument performance and the Earth's rotation, shape, and orography. The resulting product consists of geographically referenced (GeoTIFF) images, one for each of the 11 spectral bands (Table 6.3). For any TIR study, metadata files should contain information on the calibration of the instruments and the accuracy of the derived T_s values. "Ground truth" data for airborne measurements requiring adjustment for atmospheric effects can be obtained from surface-level measurements at comparison facets. More generally, the likely differences between actual and brightness surface temperatures can be evaluated by completing an inventory of the land-cover types comprising the study area, and estimating the surface emissivity at a scale that matches the TIR data.

6.4 Analysis, interpretation, and presentation

Once you have assembled the data to support your SUHI study, you will start the analysis and interpretation of those data. Here the focus is on a general approach to the examination of LST data obtained from satellite sensors, which provide an extraordinary amount of information for exploring the dynamics of the SUHI and its response to various drivers, including the physical attributes of cities (structure, cover, fabric,

Table 6.3   Summary of main advantages and disadvantages of using satellite TIR/LST data in SUHI studies.

Advantages	Disadvantages
Can provide complete (2D) spatial coverage of surface temperatures in city and surrounds.	The sensor has a near-vertical perspective of urban facets and samples horizontal roof and ground facets at the expense of vertical surfaces, such as walls.
Large community of scientists have provided standard methods for processing data, i.e., geometric and radiometric calibration.	Researcher has no control over the instrument (e.g., radiometric and spatial resolution), observation time (e.g., day or night), and the repeat cycle.
Repeated observations using the same track and orbit geometry allow examination of SUHI changes over time and seasons.	Observation data are restricted to cloud-free scenes so assessment is biased toward clear-sky conditions, making it impossible to create a SUHI climatology for all weather types.
Global coverage allows examination of SUHI properties for cities across the world. These data can be combined with other gridded data on land cover and population, for example.	There is no protocol for measuring the magnitude of the SUHI and no simple link between the SUHI and the CUHI for evaluating outdoor heat stress.
Accompanying multi-spectral data on board satellites allow estimates of type of land cover (e.g., NDVI), which can aid in SUHI interpretation.	Considerable post-processing is needed to analyze the results. This may require sophisticated software and computer resources. Moreover, detailed spatial data on urban surfaces (e.g., thermal and radiative properties of fabric) will be needed to interpret the results effectively.

and metabolism) and the geographical context (latitude and topography). However, satellite LST data also have notable disadvantages (Table 6.3).

6.4.1   Spatial analysis

GeoTIFF files of satellite data are large and require specialist remote-sensing software to manage layers of information, perform analytic operations, and generate graphical output. For general purposes, satellite imagery can be examined within Geographic Information Systems (GIS) software, such as QGIS, which is free to use. The GIS environment provides tools for the analysis of raster (cell) and vector (line) databases, and allows layers of spatial information to be integrated. Whereas the majority of thermal data are raster based, much of the information on urban infrastructure and LULC consists of points (e.g., weather station locations), lines (road networks), and polygons (building footprints).

Raster processing tools can generate statistics and histograms of cell values and identify local maxima (hot spots) and minima (cool spots). They can also define "buffers," or areas surrounding a vector feature. Fig. 6.9 illustrates a buffer of set distance

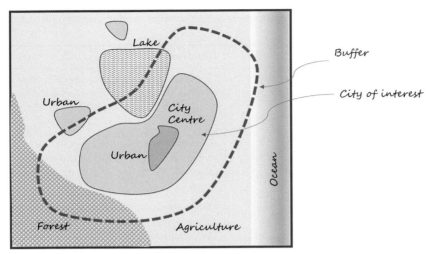

Fig. 6.9 The area surrounding a city of interest may be chosen as the benchmark for establishing the urban temperature effect. The area is indicated by a buffer that outlines the urban footprint and includes any equivalent landscape that could be urbanized, preferably at the same elevation as the city itself.

around the boundary of a city; the buffer can be used to select pixels that lie within the city's natural environs. Other filters can be used to select pixels within the buffer area that might function as a benchmark for the urban temperature effect. The same approach can be used to categorize the study area according to LULC types, such as Local Climate Zones (LCZ) (Tables 2.3 and 2.4). LCZs describe the physical character of the urban landscape based on the thermal, radiative, and geometric properties of the surface. Maps for this purpose can be generated with a variety of methods (e.g., Bechtel et al., 2015; see also Section 4.2.3.2), and then used to filter temperature values by LCZ type. The statistics associated with each type should differ significantly from each other and give basis to further analysis. Although the LCZ scheme was originally designed for CUHI work (Section 4.2.4), its underlying principles are transferable to the SUHI and provide a common framework for heat island comparisons. Other geographic information may be available that is relevant to examining the impact of the SUHI. As an example, census reports containing information on geographic attributes (e.g., housing, workplaces, population, socio-economic status) can be used to assess heat risk (Fig. 3.1). The information is compiled into maps of potential risk by grouping neighborhoods according to levels of exposure (based on population numbers), vulnerability (based on quality of housing), and hazard (extreme heat). The maps might then help to guide mitigation and adaption policies in different parts of the city.

In these types of spatial analyses, you will have to manage the various geographic coverages of the datasets. This is to ensure that information at corresponding scales can be matched. Generally, the spatial resolution of the analysis is set by the coverage with the lowest resolution. If the supplementary data are at a finer resolution than the SUHI coverage, the former will have to be aggregated to the scale of the latter. By contrast, if the supplementary data are at a coarser resolution, the SUHI data will need aggregation. There are several methods available to make these adjustments by "splitting" vector datasets based on the raster cell boundaries, and then combining the vector pieces (and their attributes) to create a new vector database that matches the SUHI raster cells. Similarly, the raster data can be converted into vector values assigned to the cell center and spatially "joined" to geographic areas with other attributes. In this way, the statistics (means and standard deviations) of Landsat 8 thermal data with a resolution of 100 m can be combined with information gathered at a much coarser scale.

6.4.2 Analysis of physical processes

The physical processes responsible for T_s variations can be examined within the context of the surface energy balance (Eq. 2.1). Warming of the urban surface is driven by the absorption of shortwave radiation ($K\downarrow$), while the temperature response depends on the transfer of available energy into the overlying air and underlying substrate. There is a considerable amount of information contained in the spectral bands acquired by EOS sensors for exploring these exchanges.

A good starting strategy is to use the spectral data available from EOS sensors to derive surface albedo and green cover. Albedo (α) is the reflectivity of the surface in the visible wavelengths ($K\uparrow/K\downarrow$), and managing it is a standard heat-mitigation policy (increasing urban albedo reduces surface heat gain). Satellite measurement of $K\uparrow$ for pixels can be compared with $K\downarrow$ (which does not vary greatly over extensive, nearly flat landscapes) to reveal areas of low reflectivity and heat gain. Indicators of green cover can also be sourced from EOS spectral data to specify surface energy transfer to the atmosphere: greater amounts of green cover are indicative of higher evaporation rates (Q_E) and less direct heating (Q_H), and thus lower T_s. The normalized difference vegetation index (NDVI) is commonly used to assess the presence of live green vegetation in a pixel:

$$NDVI = \frac{NIR - Red}{NIR + Red} \tag{6.2}$$

where Red and NIR represent the red (0.64–0.67 µm) and near-infrared (0.85–0.88 µm) wavelength bands in Landsat 8 data (Table 6.3). The NDVI ranges from -1 to 1, with positive values associated with vegetation (0.2–0.3 for shrubs and grass; >0.6 for forests). Linear regression and correlation can be used to examine the relation between T_s and NDVI, which should be strongly negative. You can then identify areas of the urban landscape where increasing vegetation might contribute to lowering T_s.

Simple urban energy balance models (e.g., SUEWS; Section 5.4.3) allow you to examine processes in some detail for selected areas of the urban landscape. These models require inputs on standard meteorological variables and the fractions of different land-cover types. Their output includes the energy balance terms and the surface (and air) temperature response. Typically, the simulated and satellite-observed temperatures are compared and statistically evaluated. As part of this process, opportunities arise to modify model parameters (within physically reasonable ranges) and improve the fit. Once satisfied with the simulation, you can extract the energy context to interpret the satellite data and its variations across the city.

6.4.3   Interpreting the results

In addition to climate modeling, fieldwork can be used to interpret the results of a SUHI study. Areas of the city that appear especially warm or cool are identified for further investigation, including a site visit. Even detailed aerial images from Google Earth can help to identify the physical character of these areas and offer insights on the processes responsible. As examples, parks that appear warm during the daytime might be experiencing water stress, while neighborhoods appearing cool might have distinct roof properties (such as white coatings or low-emissivity corrugated iron). For satellite-derived SUHI scenes, individual cells are likely to acquire a signal from a variety of surface types and, to understand the net impact, an investigation into the distribution of these types is worthwhile.

The key to interpreting results is to place the temperature data within the context of the energy exchange processes discussed in Chapter 2. This needs to account for the spatial resolution and perspective of the sensor, as well as the processes that drive the T_s response. Fig. 6.10 presents an aerial image of a neighborhood with two contrasting urban blocks: one encompasses a tree-lined park and the other a group of closely spaced buildings, alleys, and car parks. We can hypothesize on the daytime and nighttime T_s as seen from ground and airborne levels, and their response to standard

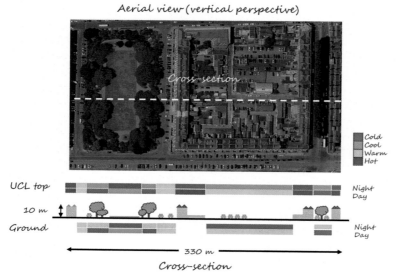

Fig. 6.10 Interpreting surface temperature at the top of the urban canopy layer (UCL), as recorded from a vertical ("bird's eye") perspective, and the corresponding ground temperature by day and night.

urban climate parameters (structure, cover, fabric) during warm weather and optimal SUHI conditions. Assuming that there is no internal heat source within the buildings and that vegetation is not water-limited, we might expect the following to occur (from a vantage point above the urban surface):

- Daytime: Roof facets are hot owing to their thin material cover (little heat-storage capacity) and lack of evaporation. Streets and paved surfaces are warm owing to their dark color and lack of water. Trees tops are cold due to evaporation. Grass-covered soils are warmed but the surface is cool due to evaporation.
- Nighttime: Roof facets are cold due to rapid depletion of stored heat and loss of longwave radiation (large SVF). Trees and paved facets close to buildings are warm due to radiation screening (small SVF). Tree tops are cool but grass surfaces are colder.

At ground level, the role of buildings and trees to limit daytime heat gain (shadowing) and restricting nighttime heat loss (screening) will be evident. Under tree canopies, the surface is cool during the day but relatively warm at night, while an above-roof perspective shows the spaces to be relatively cool by day and night. The same reasoning applies to satellite-derived TIR data, which may be relatively coarse (100–1000 m) and its T_s values integrated across many surface types and exposed horizontal facets. These include rooftops, road surfaces, and vegetated areas. Geographic databases that show the plan fractions of buildings, vegetation, impervious ground, and water are especially useful for interpreting the spatial patterns of the SUHI. Ideally, this approach would complement a surface energy balance simulation that allows one to test design ideas for mitigating the SUHI.

Field visits to relatively warm or cool neighborhoods in a SUHI map can help to interpret the SUHI magnitude and its possible impact. This gives the observational basis to identify the need and potential for heat mitigation and/or adaptation, especially in densely built areas where wall T_s values are underrepresented in satellite imagery. A simple and inexpensive hand-held TIR thermometer can be used to support the investigation. For example, a neighborhood of large storage warehouses with extensive paved areas (roads and parking) will likely appear as a hot spot in a SUHI image, but a visit may confirm that the space is lightly occupied and that the population exposure to heat stress due to the SUHI is small. Interventions for this space might therefore focus on managing urban hydrology rather than urban heat. By comparison, a densely occupied neighborhood during the day may have lower T_s but greater exposure to heat hazards. Central to this comparison is that the environment which is exposed to outdoor heat stress is located at ground level and is influenced by the 3D microscale setting, only part of which is directly observed by sensors that have near-perpendicular views. It is the roof surfaces of these settings that are overrepresented by the sensors. Clearly, cool roofs will reduce the observed SUHI and have an impact on the energy load of the building, but may have little effect on the ground-level thermal environment.

Linking satellite-derived T_s with observed near-surface T_a is not easy because the temperatures refer to different environments. Satellite T_s originate from roofs and the tops of streets and the ground, each of which contributes to the measured T_s signal based on its area-weighted (2D) longwave radiation emission (L↑). In contrast, the measured T_a responds to surface-air exchanges at surrounding facets (including walls) that shape the 3D setting. The contribution of each facet varies with wind velocity (direction and speed) and atmospheric mixing.

In a densely built area (compact midrise and highrise), the contributions of the various roof, wall, and ground facets to the TIR and air sensors will differ. Moreover, if winds are variable, the areas from which the instruments sample may not coincide— the footprint of the satellite signal is static, whereas that of the exposed thermometer within a ventilation shield is not. For this reason, the correspondence between T_s and T_a is likely to be strongest in extensive lowrise and open neighborhoods where the composition of the instrument source areas is similar, while the contribution of roofs and walls is lessened. In making this comparison, we must bear in mind the resolution of the TIR data, which typically varies from 100 to 1000 m, and the likely footprint for T_a observations, which varies with wind velocity and stability but can draw upon contributions from more than 1 km away in any given direction. This means that the relation between sets of T_s and T_a observations will depend on the weather conditions, which will modulate the air temperature footprint especially. If the comparison is based on a climatological perspective (that is, annual or seasonal), then it is important to use the meteorological circumstances associated with the contributing TIR data to select the near-surface T_a for comparison.

6.4.4   Presenting the results

In Section 5.5, we discussed various ways to present CUHI results (e.g., maps, box plots, time graphs). This discussion is certainly relevant to city-scale SUHI studies. A major distinction, of course, is that a satellite thermal image provides a complete record of the surface temperature in view, and there is little need for spatial interpolation. Exceptions occur if data are missing from pixel cells and are "filled in" with values derived from nearby pixels. Commonly, satellite-based SUHI maps depict a study area that extends beyond the case-study city, while representing the temperature in each pixel cell by assigning a color that corresponds with its value. Decisions on the type of presentation depend a great deal on the purpose of the SUHI study, whether its focus is extreme heat events or seasonal/annual variations. A robust analysis will always entail repeated and summarized observations, rather than "one-off" TIR images that may be a response to weather conditions which are not representative. However, it may be the case that, to examine specific weather events (such as heatwaves), one has limited options. If possible, an analysis of the SUHI based on single images should be placed in the context of a longer time-series that can explain observed deviations.

In all UHI studies, maps are used as a key research tool for examining patterns and generating hypotheses about causes that can be evaluated using weather and land-use/ land-cover data. Maps are also used to communicate to a wider audience the character of the SUHI, and to support explanations for heat mitigation and adaptation measures. The presentation of these data are cartographic decisions on the content and layout of the map elements, including the symbology for the temperature data, the inclusion of geographic features (roads, boundaries, etc.), and the text for labels and captions. Fundamentally, the map should be clear in its purpose and accurate in its depiction of the SUHI. As an example, Fig. 6.11 shows the average daytime and nighttime SUHI

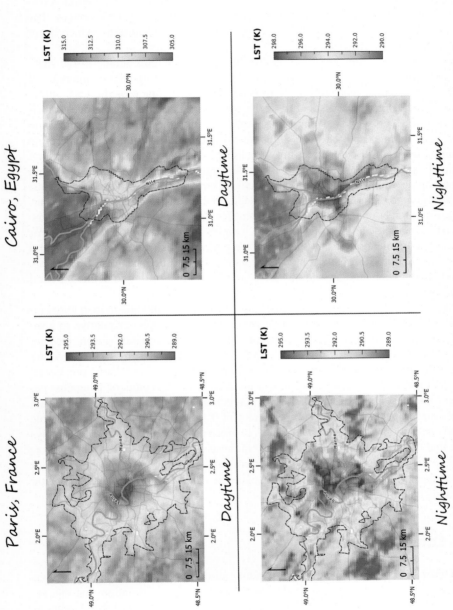

Fig. 6.11 Mean annual daytime (1030h LT) and nighttime (2230h) land surface temperature (K) over Paris (France) and Cairo (Egypt) for 2009–2013, based on data from the Terra MODIS satellite. Spatial resolution of the maps is 1 km and the official city outline and major rivers are shown.

From P. Sismanidis and B. Bechtel (Ruhr-University Bochum, Germany).

for Paris (France) and Cairo (Egypt), which have very different geographic and climatic settings. To ensure a compelling study of the SUHI in these two cities, each is presented as the mean annual land surface temperature (MAST) for the period 2009–2013, obtained as the simple average of available daily clear-sky LST images. The maps therefore represent a SUHI climatology for clear-sky conditions, which occur more frequently in warmer seasons.

Going further, the temperature legend in Fig. 6.11 fits a narrow blue-to-red spectrum for depicting relatively cool-to-warm areas, which are marked at intervals of > 1.5 K. This is consistent with the estimated nominal accuracy of the LST values in the two SUHI maps. Errors in TIR-derived T_s values are unavoidable due to variations in surface emissivity. Weng (2009) estimated that these errors can amount to 0.2–1.2 K in summer and 0.8–1.4 K in winter for the middle latitudes. It is therefore misleading to fit a full color-spectrum to marginal differences in LST. Furthermore, the spatial extent of the mapped area should be large enough to place the city within its topographic context, so that urban effects on temperature can be judged against a set background. The maps should also depict the locations of road networks, urban/metropolitan boundaries, and relevant physical features such as coastlines, rivers, lakes, and orography, all of which can help with the interpretation of T_s patterns. For effective communication at this scale, surface temperatures (rather than urban-rural surface-temperature differences, or $\Delta T_{s\ [u-r]}$) should form the basis of a SUHI map. This is to avoid confusion around the choice of background setting to define the rural environment.

To highlight neighborhood-scale variations in T_s, it is worthwhile to map temperature differences based on statistical properties (such as quartiles) or relative to a selected LULC type (using LCZs, for example). Statistical summaries (mean, median, range, etc.), histograms, and box-and-whisker plots of T_s sorted by LULC type are also helpful to assess the links between urban landscape properties (structure, cover, fabric, and metabolism) and their temperature effects. Scatter plots of T_s against parameters that describe these properties are especially effective for exploring certain controls on temperature. For example, a plot of T_s against vegetative/impermeable fraction can demonstrate the thermal effects of green cover. However, one must always ensure that the urban geographic data used in these plots are suitably scaled and formatted to support the analysis. For the purposes of providing plausible explanations based on the surface energy balance, the geographic data should be obtained at greater resolutions than the T_s data.

6.5   Concluding remarks

The SUHI is a starting point for understanding all other heat island types, yet it has historically received little attention due to the lack of instrumentation that can record surface temperatures. This has changed drastically over the last thirty years as satellite-derived thermal infrared data have become widely available. These data yield complete urban-scale 2D temperatures from standard methods that are quality controlled. However, the data are not a panacea for SUHI studies as they offer just a partial view of the 3D urban surface during clear weather only. You should therefore be well aware of how scale and perspective can affect observed temperatures and their interpretation.

A simple checklist of the steps needed to complete a successful SUHI study that adds productive knowledge to the field will correspond with those provided for the CUHI (Sections 4.3 and 5.6). It will begin with fundamental questions about what is being studied (problem), how it might be observed (resources), and for whom the results are intended (audience). If you are conducting first-hand observations, you should confirm that metadata on instrumentation, resources, and observation schedules are available; if you are using satellite TIR observations, some of these metadata are standard and the emphasis shifts to the processing, analysis, and communication of results.

1. *Establish a purpose*: Be clear on the nature of the study and the targeted audience. Is the focus on creating a climatology or examining specific events, such as extreme heat? Your response will affect decisions on the scale of the image acquisition and the averaging process.
2. *Select SUHI indicators*: Choose measures of the SUHI that are relevant to the purpose of your study. Is it to evaluate the development of the SUHI over time against a benchmark, or to examine intra-urban variations, or to identify hot spots?
3. *Gather supporting information*: Other data are needed to stratify T_s observations for analysis and interpretation of the SUHI. These include the following:
   a. Weather data for a period prior to the dates of image acquisition. The data should include information on air temperature, humidity, wind, sunshine, cloud, and precipitation. Measures of soil moisture content are especially useful.
   b. Land-use and land-cover (LULC) data to represent the urban drivers of the SUHI (i.e., surface structure, cover, fabric, and metabolism) and the topographic context of cities. These are needed to establish the natural background of the SUHI and the character of its urban neighborhoods. To aid interpretation, pay special attention to the scale of the data, which should be at a finer resolution than the SUHI data.
4. *Analyze SUHI observations*: Maps, graphs, and simple descriptive statistics can be used to isolate the controls on the SUHI and identify hot/cold spots. Correlation/regression analysis can be used to evaluate statistical relations. Urban energy balance models can be used to explore the energetic basis for SUHI variations.
5. *Communicate SUHI results*: Graphic depictions of the SUHI and its relation to urban drivers are a core part of a SUHI study. The audience should be made aware of any caveats associated with the results of a SUHI study that might limit its practical or theoretical value. In particular, nonspecialists must be advised of the distinction between surface and canopy-level UHIs, and heat mitigation and adaptation policies should be closely aligned to the UHI of interest.

References

Bechtel, B., Alexander, P.J., Böhner, J., Ching, J., Conrad, O., Feddema, J., Mills, G., See, L., Stewart, I.D., 2015. Mapping local climate zones for a worldwide database of the form and function of cities. ISPRS Int. J. Geo-Inf. 4, 199–219.

Eliasson, I., 1990/91. Urban geometry, surface temperature and air temperature. Energ. Buildings 15–16, 141–145.

Imhoff, M.L., Zhang, P., Wolfe, R.E., Bounoua, L., 2010. Remote sensing of the urban heat island effect across biomes in the continental USA. Remote Sens. Environ. 114, 504–513.

Masson, V., Gomes, L., Pigeon, G., Liousse, C., Pont, V., Lagouarde, J.P., Voogt, J., Salmond, J., Oke, T.R., Hidalgo, J., Legain, D., 2008. The Canopy and Aerosol Particles Interactions in TOulouse Urban Layer (CAPITOUL) experiment. Meteorol. Atmos. Phys. 102, 135–157.

Offerle, B., Eliasson, I., Grimmond, C.S.B., Holmer, B., 2007. Surface heating in relation to air temperature, wind and turbulence in an urban street canyon. Bound.-Layer Meteorol. 122, 273–292.

Oke, T.R., 2006. Initial guidance to obtain representative meteorological observations at urban sites. IOM Report 81, World Meteorological Organization, Geneva.

U.S. Geological Survey, 2019. Landsat 8 (L8) Data Users Handbook. Version 5. Department of the Interior, U.S. Geological Survey. https://www.usgs.gov/media/files/landsat-8-data-users-handbook.

Weng, Q., 2009. Thermal infrared remote sensing for urban climate and environmental studies: methods, applications, and trends. ISPRS J. Photogramm. Remote Sens. 64, 335–344.

Final thoughts

It has been our intention in this book to guide you through the steps of an observational UHI study. By now you will have come to a full understanding of the phenomenon, to include its energetic processes, time-space patterns, and societal implications. We have treated each of these topics with considerable detail and with reference to a large body of work in cities around the world. This might suggest to you that the study of heat islands is a mature field, and that further observation of the phenomenon is a fruitless endeavor. On the contrary, we believe that a heat island study can offer new knowledge and valuable learning experiences, but only if the study originates from a clear purpose and proceeds from a careful plan of execution.

After reading Chapters 1 through 6, you will probably have reached a decision to conduct a UHI study of your own. It is our recommendation that you carry out the study according to the guidelines described in this book. After all, the book was motivated by a great number of heat island publications that eschew the rigors of environmental science. Anyone who reads or reviews this literature will need a critical eye to separate reliable results from those that are flawed or poorly communicated. With this guidebook in hand, you are now equipped to make that separation and to embark upon a temperature study of your own design. In this chapter, we reflect on the guiding principles of such a study, helping you to avoid the most common pitfalls.

First, at the very outset of your study, it is expected that you will "know your history." By this we mean the history of heat island studies, and your familiarity with it guarantees a positive outcome. Historical awareness allows you to build upon the successes of previous work; it gives you the knowledge to ask valid questions for further research; and it leads you to useful comparisons of your work with that of others. None of these outcomes requires a comprehensive reading of the historical literature, but all of them invite a quiet perusal of the relevant works predating your own. These should include a selection of the heat island classics from the mid-20th century, and a sample of modern works from recent decades. Many such studies have been cited in the chapters of this book, while others are listed as "selected readings" at the book's end. We encourage you to browse these readings at the appropriate stages of your work.

Second, before conceiving your study, you must be confident in your understanding of the physical processes that drive the heat and energy exchanges and stores of the surface, substrate, and near-surface atmosphere. This is critical because it makes clear the fundamental distinction between heat island types, namely those of the surface, subsurface, canopy-, and boundary-layer atmospheres. Our guidebook deals mostly with the surface and canopy-layer types because these are accessible with standard thermometry and satellite imagery, and are popularly described in academic journals and the news media. Ironically, the surface and canopy-layer types are also the most readily confused, despite having very different mediums of reference—surface versus air. One must therefore be sure of this distinction before planning a study of the urban heat island—and certainly before recommending a strategy to manage its effects—as each type is associated with its own set of scales, processes, and measurement techniques. While it may seem reasonable to draw quick conclusions about the canopy-level heat island from a satellite image of surface temperatures, this is not recommended because the relation is complex and often counter-intuitive. The temptation to use satellite thermal imagery without first understanding the energetic differences between surface and canopy-layer heat islands must be resisted. In the future, this caveat will become increasingly important as access to satellite imagery improves.

Third, your choice of a UHI type to study—whether surface, canopy layer, or other—must be linked to a specific problem or curiosity that you wish to investigate. This is especially true of heat mitigation work because the thermal effects to be measured (and ultimately managed) in a city may require observation of air and/or surface temperatures. In reaching a management decision about the heat measured in a city, one must accept the fact that heat-reduction strategies should not be indicated by the heat island magnitude alone. In a hot desert city, for example, a daytime "cool island" is commonplace, giving a negative value to ΔT_{u-r}. This might be taken incorrectly to mean that heat reduction strategies are unnecessary. Researchers should instead consider the hierarchy of climatic scales that exists in a city, to include the micro, local, and regional scales. Each of these presents a range of possibilities for the thermal stresses on pedestrians and building occupants, and for the policy solutions to manage urban heat. Much of the work to identify these stresses and solutions can be done independently of the heat island magnitude (ΔT_{u-r}), but not of the heat island type, its energetic processes, or its local climates (ΔT_{LCZ}).

Fourth, the temperature data to be used in a UHI study must be collected, analyzed, and documented according to established guidelines and recommendations. These have been explained and graphically illustrated by us in this book, and are reinforced with step-by-step instructions to plan and execute a well-founded study. Here we remind you that much of the published work on heat islands has not adopted these guidelines, in part because they have been slow to develop through time. When planning your own study, keep our guidebook close at hand and be sure to ask questions related to the purpose of your study, how you might execute that study, and for whom it is intended. These questions are relevant to UHI studies of any type, but the responses will vary and will influence almost every decision to be made at subsequent stages of the work. It is therefore essential that you ponder these questions with seriousness and patience. You can then choose a methodological approach that aligns with the aims, audiences, and output of your investigation.

Upon completion of your heat island study, you may want to step back and reflect on the broader implications of what you have accomplished. If your study was education-based—e.g., as a course project or a citizen-run survey—you will have learned a great deal about designing a scientific field experiment. You might now wish to design a measurement or modeling experiment of other climatic or environmental phenomena. On the other hand, if your study was research-based—e.g., as a student thesis or a journal publication—you will have added to an existing literature on the topic, hopefully throwing new light on an important problem or question.

One such question might relate to policy options for managing atmospheric warming at the local and global scales. As with many UHI studies, yours has doubtless led to the conclusion that densely built urban areas are warmer, on average, than open areas with vegetative cover. If so, it will then occur to you that these areas of warmth are diagnostic of many other environmental consequences, such as degraded air, water, and soil quality. From this observation you could surmise that low-density development might reduce urban heating and environmental degradation at the local scale. In this assumption, however, you must recognize that low-density settlements are associated with undesirable consequences of their own (at the global scale), such as land conversion and increased energy use for transportation and building heating. If the source of that energy is carbon intensive, greenhouse gas emissions and global warming are worsened. This drives air conditioning use in buildings and the release of anthropogenic heat into the atmosphere. Ultimately, urban heating and environmental degradation are reinstated.

These and many other feedback and cause-effect relations will emerge from your work. Allow them to guide you toward new and provocative questions for deeper consideration, all of which will require a good handle on the types, causes, methods, and characteristics of urban heat islands. We hope this guidebook has given you the right tools to navigate these far-reaching questions.

Selected readings

Each of the chapters in the Guidebook is supported by a very limited list of references. Here we select a few very good works under the following themes: history and progress; classic studies; authoritative texts; surface energy budgets; heat mitigation and adaptation; modern case studies; and methodological treatments.

Arnfield, A.J., 2003. Two decades of urban climate research: A review of turbulence, exchanges of energy and water, and the urban heat island. Int. J. Climatol. 23, 1–26.

Balchin, W.G.V., Pye, N., 1947. A microclimatological investigation of Bath and the surrounding district. Q. J. Roy. Meteorol. Soc. 73, 297–323.

Bechtel, B., Demuzere, M., Mills, G., Zhan, W., Sismanidis, P., Small, C., Voogt, J., 2019. SUHI analysis using Local Climate Zones—A comparison of 50 cities. Urban Clim. 28, 100451.

Bowling S, Benson C. 1978. Study of the Subarctic Heat Island at Fairbanks, Alaska. U.S. Environmental Protection Agency, EPA-600/4-78-027, Washington, DC.

Chandler, T.J., 1965. The Climate of London. Hutchinson, London. Available at www.urban-climate.org.

Chow, W., Roth, M., 2006. Temporal dynamics of the urban heat island of Singapore. Int. J. Climatol. 26, 2243–2260.

Christen, A., Vogt, R., 2004. Energy and radiation balance of a central European city. Int. J. Climatol. 24, 1395–1421.

Duckworth, F.S., Sandberg, J.S., 1954. The effect of cities upon horizontal and vertical temperature gradients. Bull. Am. Meteorol. Soc. 35, 198–207.

Emmanuel, R., Rosenlund, H., Johansson, E., 2007. Urban shading—A design option for the tropics? A study in Colombo, Sri Lanka. Int. J. Climatol. 27, 1995–2004.

Erell, E., Pearlmutter, D., Williamson, T., 2012. Urban Microclimate: Designing the Spaces Between Buildings. Routledge, London.

Fenner, D., Meier, F., Scherer, D., Polze, A., 2014. Spatial and temporal air temperature variability in Berlin, Germany, during the years 2001–2010. Urban Clim. 10, 308–331.

Goward, S.N., 1981. The thermal behavior of urban landscapes and the urban heat island. Phys. Geogr. 2, 19–33.

Hannell, F.G., 1976. Some features of the heat island in an equatorial city. Geogr. Ann. 58 (A), 95–109.

Hawkins, T.W., Brazel, A.J., Stefanov, W.L., Bigler, W., Saffell, E.M., 2004. The role of rural variability in urban heat island determination for Phoenix, Arizona. J. Appl. Meteorol. 43, 476–486.

Howard, L., 1833. The Climate of London. Harvey and Darton, London. Available at www.urban-climate.org.

Imhoff, M.L., Zhang, P., Wolfe, R.E., Bounoua, L., 2010. Remote sensing of the urban heat island effect across biomes in the continental USA. Remote Sens. Environ. 114, 504–513.

Järvi, L., Grimmond, C.S.B., Christen, A., 2011. The surface urban energy and water balance scheme (SUEWS): Evaluation in Los Angeles and Vancouver. J. Hydrol. 411, 219–237.

Jauregui, E., 1973. The urban climate of Mexico City. Erdkunde 27, 298–307.

Klysik, K., Fortuniak, K., 1999. Temporal and spatial characteristics of the urban heat island of Lodz, Poland. Atmos. Environ. 33, 3885–3895.

Landsberg, H.E., 1970. Meteorological observations in urban areas. Meteorol. Monogr. 11, 91–99.

Li, D., Bou-Zeid, E., 2013. Synergistic interactions between urban heat islands and heat waves: The impact in cities is larger than the sum of its parts. J. Appl. Meteorol. Climatol. 52, 2051–2064.

Lowry, W.P., 1977. Empirical estimation of the urban effects on climate: A problem analysis. J. Appl. Meteorol. 16, 129–135.

Lowry, W.P., 1991. Atmospheric Ecology for Designers and Planners. Van Nostrand Reinhold, New York.

Martilli, A., Krayenhoff, E.S., Nazarian, N., 2020. Is the Urban Heat Island intensity relevant for heat mitigation studies? Urban Clim. 31, 100541.

Mills, G., 2008. Luke Howard and the climate of London. Weather 63, 153–157.

Mills, G., 2014. Urban climatology: History, status and prospects. Urban Clim. 10, 479–489.

Myrup, L.O., 1969. A numerical model of the urban heat island. J. Appl. Meteorol. 8, 908–918.

Ng, E., 2012. Towards planning and practical understanding of the need for meteorological and climatic information in the design of high-density cities: A case-based study of Hong Kong. Int. J. Climatol. 32, 582–598.

Nunez, M., Oke, T.R., 1977. The energy balance of an urban canyon. J. Appl. Meteorol. 16, 11–19.

Oke, T.R., 1976. The distinction between canopy and boundary-layer urban heat islands. Atmos. 14, 269–277.

Oke, T.R., 1982. The energetic basis of the urban heat island. Q. J. Roy. Meteorol. Soc. 108, 1–24.

Oke, T.R., 1987. Boundary Layer Climates. Routledge, London.

Oke, T.R., 2006. Initial Guidance to Obtain Representative Meteorological Observations at Urban Sites. In: IOM Report 81. World Meteorological Organization, Geneva.

Oke, T.R., 2006. Towards better communication in urban climate. Theor. Appl. Climatol. 84, 179–190.

Oke, T.R., Johnson, G.T., Steyn, D.G., Watson, I.D., 1991. Simulation of surface urban heat islands under 'ideal' conditions at night—Part 2: Diagnosis of causation. Bound.-Lay. Meteorol. 56, 339–358.

Oke, T.R., Spronken-Smith, R.A., Jauregui, E., Grimmond, C.S.B., 1999. The energy balance of central Mexico City during the dry season. Atmos. Environ. 33, 3919–3930.

Oke, T.R., Mills, G., Christen, A., Voogt, J.A., 2017. Urban Climates. Cambridge University Press, Cambridge, UK.

Peterson, T.C., Owen, T.W., 2005. Urban heat island assessment: Metadata are important. J. Climate 18, 2637–2646.

Roth, M., Oke, T.R., Emery, W.J., 1989. Satellite-derived urban heat island from three coastal cities and the utilization of such data in urban climatology. Int. J. Remote Sens. 10, 1699–1720.

Sakakibara, Y., Matsui, E., 2005. Relation between heat island intensity and city size indices/canopy characteristics in settlements of Nagano Basin, Japan. Geograph. Rev. Jpn. 78, 812–824.

Skarbit, N., Stewart, I.D., Unger, J., Gal, T., 2017. Employing an urban meteorological network to monitor air temperature conditions in the 'local climate zones' of Szeged, Hungary. Int. J. Climatol. 37 (S1), 582–596.

Stewart, I.D., 2011. A systematic review and scientific critique of methodology in modern urban heat island literature. Int. J. Climatol. 31, 200–217.

Stewart, I.D., 2019. Why should urban heat island researchers study history? Urban Clim. 30, 100484.

Stewart, I.D., Oke, T.R., 2012. Local Climate Zones for urban temperature studies. Bull. Am. Meteorol. Soc. 93, 1879–1900.

Sundborg, Å., 1950. Local climatological studies of the temperature conditions in an urban area. Tellus 2, 222–231.

Szymanowski, M., 2005. Interactions between thermal advection in frontal zones and the urban heat island of Wrocław, Poland. Theor. Appl. Climatol. 82, 207–224.

Terjung, W.H., O'Rourke, P.A., 1980. Simulating the causal elements of urban heat islands. Bound.-Lay. Meteorol. 19, 93–118.

Voogt, J.A., Oke, T.R., 1998. Effects of urban surface geometry on remotely sensed surface temperature. Int. J. Climatol. 19, 895–920.

Voogt, J.A., Oke, T.R., 2003. Thermal remote sensing of urban climates. Remote Sens. Environ. 86, 370–384.

Wanner, H., Filliger, P., 1989. Orographic influence on urban climate. Weather Clim. 9, 22–28.

Yow, D.M., Carbone, G.J., 2006. The urban heat island and local temperature variations in Orlando, Florida. Southeast. Geograph. 46, 297–321.

Index

Note: Page numbers followed by *f* indicate figures and *t* indicate tables.

Printed in the United States
by Baker & Taylor Publisher Services